HANDBOOK OF
RESEARCH
LABORATORY
MANAGEMENT

HANDBOOK OF RESEARCH LABORATORY MANAGEMENT

VIRGINIA P. WHITE

iSi PRESS®
Philadelphia

Published by

iSi PRESS®

3501 Market Street, Philadelphia, PA 19104 U.S.A.
(215) 386-0100, Cable: SCINFO

Library of Congress Cataloging-in-Publication Data

White, Virginia P.
  Handbook of research laboratory management.

  Includes bibliographies and index.
    1. Laboratories—Management—Handbooks, manuals,
etc.   I. Title.
Q183.A1W47   1988   507′.2   87-22606
ISBN 0-89495-065-7

Printed in the United States of America.

To Mary Esther Gaulden
who introduced me to the world of science

and

To the memory of Alexander Hollaender
who gave me the opportunity to be
a part of that world.

ACKNOWLEDGEMENTS

My orientation to the world of science began at the Biology Division of the Oak Ridge National Laboratory in Tennessee in 1955. As an administrator, my role there was never easy but it was always challenging and exciting, thanks to the high quality of the scientific staff and the research they produced under the leadership of Alexander Hollaender. From the time I arrived until I left in 1967, I was considerably aided in fulfilling my responsibilities by the encouragement and cooperation of the staff, including technicians, craftspeople, animal caretakers, clerks, and other administrators. I am especially grateful to the scientists who included me in scientific discussions and talked with me personally and at length about their work even when I could not fully understand the details or grasp the implications. It was through these associations that I gained insight into the needs and aspirations of scientists and some comprehension of the stress they undergo in their search for answers that may be obscure, elusive, or just not there.

Those to whom I am most grateful for their contribution to my education are: Howard I. Adler, W. A. Arnold, John Cook, Fred de Serres, Mary Esther Gaulden, Ed Grell, Rhoda Grell, John Jagger, Dan L. Lindsley, Jane K. Setlow, Richard B. Setlow, Dorothy Skinner, Arthur C. Upton, R. C. (Jack) von Borstel, Sheldon Wolff, and S. F. Carson, who is now deceased.

Between 1956 and 1967, approximately 150 scientists from thirty countries outside the United States worked in the Biology Division under the Visitors Exchange Program, which was part of President Eisenhower's "Atoms for Peace" effort. They came from Latin American countries, from Europe, Canada, Australia, and the Mid- and Far East and ranged from young postdoctoral fellows to eminent scientists. Some stayed for two or three years; others were short-term visitors. Many internationally known scientists came to lecture and see the work of the laboratory under sponsorship of their own governments or other sources. From those I knew well, I learned about scientific organization and philosophies that exist in other parts of the world—a perspective that I could have received in no other way. Those I remember with special gratitude for acquainting me with the international world of science are: Douglas Davidson (England), Sohei Kondo (Japan), Heinrich Kroeger (Germany), Raymond Latarjet (France), Sohan P. Modak (India), Amir Muhammed (Pakistan), Crodowaldo Pavan (Brazil), and Vu-Thi-Suu (South Vietnam). Adriano Buzzati-Traverso (Italy) and Tuneo Yamada (Japan), both now deceased, were also extremely helpful.

At the Salk Institute where I went in 1968, I was affiliated with outstanding scientists, a number of whom were generous in spending time talking with me about their aspirations, frustrations, and reactions to successes and failures. Jonas Salk, founding director, was exceptionally candid and forthcoming in his conversations. Other scientists there to whom I feel indebted for continuing my education are: Renato Dulbecco, Walter Eckhart, Robert W. Holley, and especially Jacob Bronowski, whom I later visited in England when he was filming the remarkable *Ascent of Man* series. It was a great loss when Dr. Bronowski died in 1974.

Nonscientist administrators at the Salk Institute from whom I also learned much are Joseph Slater, John Hunt, and Lorraine Friedman.

Over a period of several years, in order to gain information for this book, I visited laboratories in the United States and England where I talked with scientists, administrators, and those who were both. I would like to acknowledge especially the courtesy and kindness of those I met in England who generously allowed me to visit their laboratories, to speak with them at length, and who supplied me with useful information and documents that have been referred to in various places throughout the book. They were: Sir Brian Pippard, Sir Nevill Mott, and J. S. Mitchell at Cambridge University; John R. Vane and W. Dench at the Beckenham Laboratories of the Burroughs Wellcome Company; Sir Brian Flowers, Rector of London's Imperial College of Science and Technology (currently Lord Flowers, Vice Chancellor of the University of London); M. G. P. Stoker, Imperial Cancer Research Fund, London; Joseph Rotblat, The Medical College of St. Bartholomew's Hospital, London; and J. W. Boag, Royal Marsden Hospital, Sutton.

Arthur D. Rörsch of the European Molecular Biology Organization (EMBO) and Leiden State University, The Netherlands, talked with me when he was visiting in the United States, and he later sent me very informative material concerning international cooperation in scientific research.

In the United States, I would like to cite particularly the following institutions and libraries I visited and people I talked with:

Institutions:

Brookhaven National Laboratory, Upton, New York: R. Christian Anderson, W. R. Casey, Gerhart Friedlander, Maurice Goldhaber, Bernard Manowitz, Eliot Shaw, and Harold Siegelman. Also Leland Haworth and George Vineyard, who have since died.

Burroughs Wellcome Company, Research Triangle Park, North Carolina: R. M. Cresswell, Pedro Cuatrecasas, Iris B. Evans, George H. Hitchings, Fred C. Kull, and Robert A. Maxwell.

University of California at San Francisco: Harvey Patt (deceased) and Sheldon Wolff.

Cold Spring Harbor Laboratory, Long Island, New York: James D. Watson.

Institute for Scientific Information, Philadelphia, Pennsylvania: Calvin Mark Lee.

IBM Thomas J. Watson Research Center, Yorktown Heights, New York: Louis Branscomb, L. A. Cookman, and Antoinette A. Emerson. Tom Murphy supplied me with very informative material.

National Institute of Environmental Health Sciences (NIEHS), Research Triangle Park, North Carolina: David P. Rall, Frederick J. de Serres, and Paul Waugaman.

Oak Ridge National Laboratory, Oak Ridge, Tennessee: Herman Postma, C. R. Richmond, and John B. Storer. Those who were especially helpful at ORNL were: E. J. Frederick, Mary Jane Loop, and Esther H. Thompson.

Rockefeller University, New York, New York: Rodney W. Nichols.

Scripps Clinic and Research Foundation, La Jolla, California: Frank J. Dixon and Robert J. Erra.

Smithsonian Institution, Washington, D.C.: Wilton S. Dillon.

University of Texas at Dallas, Texas: John Jagger, Stuart C. Fallis, J. G. Moore, and C. S. Rupert.

University of Texas Health Science Center at Dallas, Texas: Mary Esther Gaulden, Jean K. Miller, Gerald Mussey, Johnnie Reynolds, Charles C. Sprague, Julius E. Weeks, and Katherine L. Chapman, who was most helpful to me in obtaining useful material.

Libraries:

Library of Congress, Washington, D.C.: Dan Melnick.

Memorial Sloan-Kettering Cancer Center Library, New York, New York: June Rosenberg.

New York Academy of Medicine Library, New York, New York: Brett Kirkpatrick and Art Downing.

People in many places went out of their way to help me obtain information and documentation and I can mention only a few of them: Marion Ball, Director of Computing Services, University of Maryland School of Medicine, Baltimore; Rod Cochran, State University of New York, Syracuse; Marianne L. Dillon, Battelle Development Corporation, Columbus,

Ohio; Bob Freeman, Director, East Tennessee Development District; William J. Gartland, National Institute of Allergy and Infectious Diseases, NIH, Bethesda, Maryland; Norman Latker, Patent Counsel, U.S. Health and Human Services Department; Michael Loop, Medical Science Center, University of Alabama–Birmingham; Richard T. Loutit, National Science Foundation; Mark Meade, Robert Ubell Associates, New York City; Ronald O. Rahn, School of Public Health, University of Alabama–Birmingham; Robert M. Simon, National Research Council; Herbert P. Tinning, American Society of Mechanical Engineering (ASME); Martha E. Williams, University of Illinois–Urbana/Champaign; and George Wise, Historian of General Electric Company.

Portions of the manuscript were read by Richard B. Setlow, Brookhaven National Laboratory and by George N. Eaves, National Institutes of Health, Bethesda, Maryland, who have my undying gratitude for their insightful and intelligent suggestions, which were exceedingly beneficial.

The production process involved numerous people representing many skills and talents. I particularly want to thank Tom Repensek for the intelligence and meticulous care he brought to the laborious copyediting task; Elaine Cacciarelli and Anne Just for editorial assistance; Rudi Wolff for his talented design; and Carol Ingram who typed the manuscript mostly from dictated tapes.

Acknowledgements by authors usually show very little originality and one phrase appears so often that it may sound perfunctory—that which thanks someone "without whom this book would never have been written." In my case, the citation is far from perfunctory—my gratitude to Robert Ubell is profound. I can truly say that without him this book *would not* have been written.

Virginia P. White

New York City
October, 1987

INTRODUCTION

Scientific organizations—research laboratories and R&D groups—are different from other enterprises, and in significant ways they are different from each other. What they have in common is an ancestry that includes the scholarly society and the factory—the gifted inventor and the theoretical dreamer.

This book begins by tracing the history of the rise and development of research groups, noting how their diverse antecedents produced the unique institution that is the modern laboratory. The size, purpose, and research focus, as well as the social and political atmosphere in which it operates, all influence the management principles and style of a laboratory. A major determinant of its success is leadership. The selection of the right person at the right time in the life of the organization can be decisive for achieving honor and distinction.

There is no one type of person or one cluster of abilities that will ensure successful leadership in all scientific groups or even in one particular kind of laboratory. Descriptions of acclaimed scientific administrators reveal that they emerge from widely varying backgrounds and traditions and hold different philosophies. The key to their success seems to lie in two specific factors: One is the ability to interweave their distinctive talents with the resources and objectives of the institution in pursuit of its highest level of achievement. The other is knowledge.

The scientist who rises to the head of a group or who is plucked out of the laboratory and given a leadership role is, in nearly all cases, highly knowledgeable in one or more areas of science. Those whose work has been recognized by honors and awards or by election to membership in highly regarded scientific societies gain the respect and admiration of their colleagues, an essential element in exerting leadership. But scientific knowledge and talent alone do not ensure success in leading a scientific organization. Another kind of knowledge is also required—less esoteric, perhaps, than science and certainly more obtainable than natural talent. It is that area of knowledge to which this book addresses itself.

The head of a laboratory must be able to:

- assemble and maintain a staff capable of advancing the work to which the laboratory is dedicated, and to see that the results of that work are disseminated to the scientific community and to the public;
- acquire and manage the financial resources so the quality and quantity of yield (the "product" in relation to costs) bring credit to the laboratory and to the organization of which it is a part;
- make continuing education an integral part of every aspect of the research program, not only for scientific personnel but for the entire staff, the community, and the younger generation of scientists or potential scientists.

In order to meet this challenge, the leader of a research organization, whatever its size, needs to be informed in some areas that most scientists probably never even think about—personnel selection and administration, acquisition and management of funding, and internal and external communications. For the first time, in many cases, these former independent researchers are now forced to think about remuneration and rewards to other scientists; legal aspects of patents and copyrighted material; protection of laboratory personnel and of human and animal subjects from hazards in connection with work going on in every part of the laboratory.

The rewards of the successful scientific administrator are different from those accorded to the individual researcher. One scientist named as director of a famous laboratory described it this way: "I view 'this as another phase in my career in which I work toward encouraging others and providing the atmosphere and resources for a number of scientific activities. It is a different achievement from individual and independent research, but very interesting and rewarding."

Although crucial to the development of a successful research laboratory, inspired leadership alone is not enough. The most gifted leader can be thwarted by the insensitivity or incomprehension of those administrators in the upper echelons of the institution of which the laboratory is a part and on which it depends for support.

Very few research organizations stand alone and answer only to a governing board. They are usually one unit of a larger enterprise and must fit into a superstructure, conforming to the policies and procedures established by higher authority. Colleges and universities, corporations, and national, state, and local government agencies provide the settings for most research activities. The heads of those institutions may be scientists, but as a rule they are not, and it is they who must be especially sensitive to the duality that exists in modern research—the need of scientists for maximum creative freedom, linked to the need of the institution to meet its established goals—the scholarly society versus the factory. When conflicts occur, scientists are likely to be adamant in demanding freedom to pursue their ideas. And even those administrators who may be sympathetic to that view are faced with the demand to balance it with the institution's needs. Students must be taught and trained; new products must be developed; and the goals of the sponsor who provides support for the institution and the laboratory must be pursued.

The responsibility for mediating such conflicts falls into the hands of the head of the laboratory, who must be able to negotiate the inevitable compromises leading to amicable arrangements. A natural talent for diplomacy helps, but some of these skills can be learned.

In order to maintain the morale of the research staff and also meet the goals of the institution, it is essential that administrators in the upper echelons and the bench scientists understand each other's objectives and points of view and have some insight into each other's problems. This seldom happens by chance. People tend to spend time with colleagues whose attitudes and interests are similar to their own. The size of the institution and the location of the laboratory within it will determine what kind of interchanges are possible, but the laboratory leader is always challenged to come up with ingenious mechanisms for bettering communication between the two groups.

Institutional executives should have a clearly expressed standing invitation to attend laboratory seminars and, on occasion, special invitations should be issued for those in which they might have a particular interest. They might also be invited to research staff meetings and from time to time be asked to speak to the group about the goals of the institution and to answer questions on matters of concern to the research staff. For their part, the institutional directorate might arrange for scientists to attend some meetings held primarily for discussion of administrative or business matters. Scientists might also be invited to hear talks relating to the growth and development of the institution as a whole. They may be invited occasionally to departmental and other administrative meetings normally attended only by the laboratory head.

It is a good idea also to create opportunities for the research staff and administrative executives to participate jointly in ceremonies of interest primarily to the laboratory. For major events, such as a scientist receiving a particularly high honor—a prestigious award, election to the National Academy of Sciences or to an important office in a scientific society—the institution should take note by organizing special ceremonies to which everyone is invited.

The laboratory head must take every opportunity to acquaint the institution's administrators with the level and kind of support services needed by a fully functioning research staff of high quality. Educating the upper echelon to this necessity should not be embarked upon in conjunction with the annual budget presentation, but should be a process that goes on throughout the year.

A chemistry lab in a university, an industrial R&D group, a pharmaceutical research company, and the large national laboratories have widely different organizational and management problems. But all of them have a need for information in the basic areas of personnel, funding, safety, buildings, equipment, research services, communications, and the provision of information sources. This book offers guidance in these areas that is applicable to some degree to scientific organizations of any size, operating in a modern social system. It should be useful not only to laboratory directors but also to institutional executives whose responsibilities involve research. It may also be enlightening to bench scientists, as it defines the concerns of those responsible for research management and direction. Even those staff members whose activities are only peripherally or laterally concerned with research will find certain sections informative. The subject of external funding, for example, will be of interest to the university director of development; the corporate executive concerned mainly with product management may find the section on patents informative. The plant engineer in any institution of any size needs information on the provision of facilities and equipment and their maintenance.

The work emerging from research laboratories has changed and continues to change our view of the world and our place in it. Activity of that importance deserves to be organized and managed with all the knowledge and skill available. Institutions that sponsor the operation of research programs, and the individuals who take on the task of heading research laboratories, have an awesome responsibility. To produce the highest quality of research in the most efficient way is one of the greatest challenges a scientist can face. But it is well worth the doing. The personal rewards, great as they are, are insignificant compared with the potential for the human race.

CONTENTS

CHARTS, LISTS, GRAPHS

Chapter 1

THE DUALITY OF THE MODERN RESEARCH LABORATORY

The combined essences of heaven and earth became the yin and yang . . .

—Huai-nan Tzu, Anonymous work
compiled at the court of Liu An, Second Century B.C.

Scientists and scholars have for centuries sought out their peers in order to exchange ideas with others of equivalent learning and interests, but it was not until late in the Industrial Revolution that the knowledge and talents of a number of scientists began to be assembled under one roof to attack problems requiring a wide range of techniques and capabilities.

This development resulted in a strange union that produced the institution we know as the modern research laboratory—the progeny produced by the mating of the scholarly academy and industry. With a background that includes two such inharmonious forebears, it is not surprising that research institutions are awkward to organize, difficult to work in, and, some say, impossible to manage.

NINETEENTH-CENTURY RESEARCH

By the middle of the nineteenth century, improved ease of travel and communication made it possible to learn about work going on in other cities and even other countries and eventually to launch cooperative studies between widely separated researchers. By that time, too, the body of scientific knowledge was growing so fast that no one person could keep up with all of it, so specialization and thus mutual dependency developed among scientists. At the same time, expensive and complicated instruments made it possible to complete in a matter of weeks a task that formerly took months or even years.

The invention and standardization of scientific apparatus, too expensive for the individual researcher to buy and too complicated for one person to build or even operate, contributed to the decline of the home-based laboratory and to the formation of clusters of researchers—the modern laboratory.

The role of machinery and equipment probably accounts for the fact that some of the world's greatest laboratories were spawned within the industrial community. The Carlsberg Laboratory of Denmark, for example, which has produced a body of elegant fundamental research, was founded by John Christian Jacobsen only partly for "pure" research. One of its original mandates was to study those branches of science that are particularly important for the processes of malting, brewing, and fermentation. Jacobsen wanted to make a better beer.

Another product-oriented institution that has made significant contributions to basic research is the Burroughs Wellcome Company. Originally named Burroughs Wellcome and Co., it was formed in London in 1880 by two Americans who had been fellow students at Philadelphia College of Pharmacy, Silas M. Burroughs and Henry S. Wellcome. The company pioneered new high-quality drugs with scientifically precise dosages and has achieved a distinguished record creating improved pharmaceuticals for human and animal use.

The Wellcome Research Laboratories were set up in London in 1894. When Burroughs died in 1896, Wellcome took over sole ownership and he began to emphasize scientific research to an extent that had no precedent in the commercial world. Before accepting a research post there in the early 1900s, Sir Henry Dale hesitated, fearing that he would "be selling my scientific birthright for a mess of commercial pottage." When he consulted friends about it, he said, "they unanimously advised me to have nothing to do with it." Sir Henry later became a member of the Order of Merit, president of the Royal Society, and in 1936 shared a Nobel Prize with Otto Loewi of Graz University, Austria, for their discoveries relating to chemical transmission of nerve impulses.

We have Wellcome Research Laboratories to thank for the word "tabloid," which it coined for its line of compressed drugs; the word "tabloid" has been used since then for a newspaper or for anything that is condensed or compressed. Wellcome's Tabloid Medical Chests went up the Amazon with Theodore Roosevelt and were part of the expedition that first scaled Mount Everest. They were carried by Shackleton, Scott, Amundsen, and other polar explorers. The Wellcome Tropical Research Laboratories, which included a floating laboratory on the Upper Nile, were opened in Khartoum in 1901. By 1935, when the Sudan government closed down the labs, Khartoum (where most of the natives had suffered from malaria in 1901) was one of the healthiest cities in Africa.

Wellcome died in 1936 and left the ownership of the company to the Wellcome Trust, with instructions that all profits distributed as dividends be applied to the support of medical and allied research throughout the world. The company, now the Wellcome Foundation, employs over 18,600 people worldwide and spends about $96.6 million on research and development in its laboratories each year. The Wellcome Trust distributes the dividends from the Wellcome Foundation for medical and allied research throughout the world and, since 1936, has given more than $125 million to support individuals, universities, and the Zoological Society of London.

While the American Wellcome Research Laboratories at Triangle Park are product-oriented—with the ultimate objective to develop disease-combatting pharmaceutical compounds—there is great emphasis on basic research. Scientists at the Triangle Park laboratory have beautiful facilities that provide ample equipment and support services. Many scientists there report that they spend as much as 50 percent of their time on basic research. They appear to have no more pressure to produce work that will lead to a patentable product than university research professors have to publish—perhaps not even as much.

During the latter half of the nineteenth century, as Burroughs and Wellcome were setting up their pharmaceutical lab in London and Jacobsen began research on fermentation processes in Copenhagen, Thomas A. Edison was establishing his laboratory in Menlo Park, New Jersey, to develop the electric light.

Ultimately, the Edison Laboratory became one of the parents of the General Electric Company—the first industrial laboratory in the United States to cultivate fundamental research. Although GE was concerned mainly with extensive electrification of industry, its laboratories performing research in physics, metallurgy, and chemistry contributed to fields not directly allied with the electrical industry.

The Cavendish Laboratory, opened in 1871, is considered by many as the jewel of physics labs. Not in an industrial setting, it is at one of the world's great centers of learning—Cambridge University. Yet in his history of the laboratory, J. G. Crowther writes, "The early years in the Cavendish Laboratory correspond, in a way, with those in the Boulton factory at the beginning of the Industrial Revolution,"[1] when Matthew Boulton and James Watt made the steam engine an efficient means of driving machinery.

THE CLASSICAL PERIOD

Long before any of the inventions that led to machinery production and labor-saving devices, long before the Industrial Revolution—as far back as the third century B.C.—scientists had begun to meet with and to have their work scrutinized and evaluated by other scholars of similar knowledge and interests.

Around 387 B.C., Plato founded an academy for the systematic pursuit of philosophical and scientific research, and during his life it was a recognized authority in mathematics and jurisprudence. All of the most important mathematical work of that century was done by friends or pupils of Academy member Plato. Theaetetus was the founder of solid geometry. Archytos, the inventor of mechanical science, was a friend and correspondent, and Speusippus, Plato's nephew and successor, was a voluminous writer on natural history. Aristotle's biological works belong to the period in his career immediately after Plato's death and before he broke with the Academy. The creation of Plato's Academy as a permanent society for the pursuit of the exact and humane sciences has been viewed by some as the first university.

Also in the third century B.C., Alexander the Great's trusted general, Ptolemy I, a member of his elite bodyguard who later became satrap of Egypt upon Alexander's death, founded an academy at Alexandria where scholars gathered and studied classical sciences. That academy, which might also lay claim to being the first university, formed the nucleus of the famous library of Alexandria, which became the intellectual center of Hellenistic culture under Ptolemy Philadelphus (Ptolemy II).

In the first century B.C., Cicero belonged to the New Academy which grew out of Platonism but rejected the sensationalist dogmatism of the Stoics. Its principles permitted followers to hear all arguments for and against a proposition and to accept the conclusion which for the moment appeared most probable. Cicero's *Academica* is the literary record of the quarrel between the New Academy president, Philo of Larissa, and his scholar, Antiochus of Ascalon. There was a gradual rapprochement between Stoicism and the tenets of the Academy, but the writings of Plutarch in the fourth century A.D. are the best example of the popular Platonism or Neo-Platonism of the New Academy.

THE BIRTH OF SCIENTIFIC ACADEMIES

Academies of various kinds existed in Italy during the Middle Ages; one of the most famous was the Academia Platonica, founded about 1442 by Cosimo de' Medici (Cosimo the Elder). As its name implies, it was chiefly occupied with Platonic studies, but also included work on Dante and the purification of the Italian language. Throughout the Renaissance the Medici family was famed for its support of art and learning. In 1657, Leopold de' Medici, who had been a student of Galileo, founded the Academia del Cimento at Florence where nine scientists united their experimental efforts and published their results jointly. This seems to be the first organized experimental "laboratory" on record, but it lasted only ten years.

The old Académie des sciences in France, established by a grant from King Louis XIV in 1666, at first was more of a laboratory and observatory than an academy. Before its formal organization, a number of men of science had been meeting together for some thirty years—including Descartes, Gassendi, and Blaise and Etienne Pascal. On his first visit to Paris in 1640, Hobbes met with this group. Chemists, physicians, anatomists, and mathematicians formed the nucleus of the formal academy. Christiaan Huygens and Bernard Frenicle de Bessy were members as were other distinguished scholars. Several foreign notables were also invited to join as associate members, among them Sir Isaac Newton, the Danish astronomer Ole Roemer, and the German physician and geometer Ehrenfried Tschirnhaus.

The labors of the academicians were diverted from the pursuit of pure science to such works as the construction of fountains and cascades at Versailles, and the mathematicians were employed to calculate the odds of the games of *lansquenet* and *basset*.

In 1699, the Académie des sciences was reconstituted and by its new structure consisted of twenty-five members: ten honorary members of high rank interested in science, and fifteen pensioners, the working members, who numbered among them three geometricians, three astronomers, three mechanicians, three anatomists, and three chemists. The leading spirits of the academy during that period were Alexis Clairaut, René-Antoine Réaumur, and Bernard Fontenelle, the popularizer of science. Prominent members during the eighteenth century were Pierre-Simon, Marquis de Laplace, George-Louis Buffon, Joseph-Louis Lagrange, Jean d'Alembert, Antoine-Laurent Lavoisier, and Bernard de Jussieu, the father of modern botany. The academy was suppressed in 1793, as were all French academies, including the Académie française, established in 1635, which began working on a dictionary of the French language in 1639. In 1795, the Institut national was established to replace all academies. By 1816, the Académie des sciences was reconstituted as a branch of the Institut.

Surprising as it seems today, the universities contributed very little to the advance of science before the nineteenth century. In Germany, as in most of Europe, research laboratories were not considered a proper part of the university's facilities; institutes were established just outside the university structure to separate individual professors from teaching and administrative duties. The prevailing belief was that the university was concerned with literary, classical, or liberal studies, and did not include science or technology.

Vestiges of this school of thought were evident in the negotiations that preceded the establishment of the United Nations Educational, Scientific, and Cultural Organization (UNESCO) in 1945. Committeemen of the classical humanist tradition thought it better to confine their efforts to education and culture, arguing that only those fields were concerned with the preservation and development of moral values. Except for the presence on the committee of Julian Huxley, for whom science was the most advanced form of culture and the basis of all values, the "S" might never have appeared in UNESCO.

The scientific societies that sprang up within or around the universities were extracurricular and functioned somewhat like private clubs. One group, which may have been the first of its kind in Europe, met regularly in the home of William Gilbert, physician to both Elizabeth I and James I. Its membership was composed of scholars interested in the secrets of nature and the art of their discovery through the experimental method. Among the scientific results discussed were such landmarks as Johann Kepler's *New Astronomy* and Sir William Harvey's work on the circulation of the blood.

Sir Francis Bacon, who was Lord Chancellor of England from 1618 to 1621 as well as a scientist and philosopher of empiricism, was not a member of this group, and Harvey is reported to have said that Bacon "writes philosophy (i.e. science) like a Lord Chancellor." Nevertheless, it was Bacon who inspired at least two learned societies, giving credence to his claim that he "rang the bell which called the wits together." When the Royal Society of London for Improving Natural Knowledge began to hold its weekly meetings in 1645, almost twenty years after Bacon's death, it was described as a meeting of "divers worthy persons, inquisitive into natural philosophy and other parts of human learning, and particularly of what hath been called the *New Philosophy of Experimental Philosophy.*" Some of those "worthy persons" formed yet another group in 1647, at Oxford, called the Philosophical Society of Oxford, that met regularly in the rooms of John Wilkins, warden of Wadham College, but ultimately moved to Gresham College in London.

In 1742 the Royal Danish Academy of Sciences and Letters was founded for the advancement of mathematics, astronomy, natural philosophy, and other disciplines, by the publication of scientific works and by the endowment of research.

RESEARCH IN THE UNITED STATES

When the wits were first called together in the United States, it was not to form a gentlemen's club or to create a scientific society. It was to achieve a purpose, to perform a function. Within that circumstance lies a clue, perhaps, to the future development of science and scientific institutions in America. In 1743, the American Philosophical Society Held at Philadelphia for Promoting Useful Knowledge, later called the American Philosophical Society, was founded as a result of Benjamin Franklin's treatise, "A Proposal for Promoting Useful Knowledge Among the British Plantations in the United States."

"Useful knowledge" was also what Congress sought to promote in passing the Morrill Act in 1862. Commonly known as the Land Grant Act, it called for the establishment of at least one institution in each state of the Union in which scientific and technical studies were to be given equal rank with classical learning and education in the professions.

It was also what President Abraham Lincoln had in mind in 1863 when he established the National Academy of Sciences to "investigate, examine, experiment and report upon any subject of science or art desired by any department of government." President Woodrow Wilson, during World War I, implemented the purpose of the National Academy by establishing the National Research Council, at least partly for the purpose of developing the United States as a military power. It is significant that seventeen years earlier, in 1846, the Smithsonian Institution was established in the United States for the "increase and diffusion of knowledge among men" with no reference to usefulness. It resulted from the bequest of James B. Smithson, an Englishman, who had never even visited the United States and whose eccentricity in making such a bequest raised questions in some circles regarding his sanity.

Throughout the nineteenth century the United States was a developing nation concerned with exploring and settling the frontier, building transportation and communication lines to reach the farthest outposts, and stimulating industrial growth. In the first half of the twentieth century, industry moved into mass production and mass marketing, and the technologies dealing with those activities received major attention. Little or no public attention was paid to fundamental research or basic studies. Like developing countries today, the United States imported basic knowledge from abroad. Until the very eve of World War II, the United States was dependent upon European discoveries in basic science which, through the application of what came to be known as "Yankee ingenuity," were exploited to develop natural resources, create jobs, raise the standard of living, and establish the United States as a world leader in the new technological society.

When the devastation of World War II all but wiped out scientific research in Europe, the United States had already acquired a cadre of scientific talent as a result of the emigration of Jews from Germany and the Nazi-dominated Central European countries during the 1930s. Under the pressure created by the exigencies of the war, and aided by the brilliance and talent of newly arrived scientists, the United States launched an intensive program of scientific research and rapidly became the leading source of new knowledge in several fields, most notably in atomic energy.

By 1944, President Roosevelt was already concerned with establishing a plan for maintaining the momentum that had been reached in scientific research during the war. He turned for such a plan to Vannevar Bush, under whose leadership, as director of the Office of Scientific Research and Development, this momentum had been achieved by an unprecedented mobilization of the research talent within the nation's universities. In *Endless Horizons*, Bush summarized his thoughts about such a plan:

> During the war years we drew heavily on our scientific capital, making great advances in applied science—in radar, rockets, antiaircraft gunnery, in immediate therapeutics, in transportation. To do so, we had to give up fundamental research and so we had to sacrifice the future to the present. We must now replenish the reservoirs of fundamental knowledge. It is through the application of the results of vigorous fundamental research that we have in the last years created extensive industries, secured productive employment for our people, raised our standard of living and of general education, and increased the national income upon which government draws for the general good. We must be able to rely in the future on fundamental science to provide the basis for these things to a greater extent than we have in the past.[2]

A key provision in the Bush recommendation was that public and private colleges, universities, and research institutes should be in full control of policy, personnel, and the method and scope of the basic research program, even though its support would come from the proposed government agency.

But, because of the government's inevitable need for accountability, a further provision stated:

> While assuring complete independence and freedom for the nature, scope and methodology of research carried on in the institutions . . . the usual controls of audits, reports, budgeting, and the like should, of course, apply to the administrative and fiscal operations of the Foundation.[3]

"The Foundation" was the proposed institution that eventually emerged as the National Science Foundation.

The obvious conflict implied in those two provisions sums up succinctly the atmosphere in which the head of a research laboratory functions, regardless of the source of the funds: while striving for complete independence and freedom for the nature, scope, and methodology of research carried on in the institution, the usual controls of audits, reports, budgeting, and so on must be imposed. There is no escape. While fostering the spirit of unfettered imaginative inquiry within the lab, one is always mindful that the heights to which creativity can soar are determined by such practical considerations as space, equipment, tools, personnel, and—an indispensable element in the "unfettering" process—a livelihood for the researcher.

The extraordinary vision and wisdom of Bush's recommendations can be credited in large measure with raising the scientific stature of the United States to the first rank. That stature has been recognized in many ways, but, in view of the number of Nobel awards regularly conferred upon the United States' scientists, no one can question the change that has occurred since 1946. It is interesting to compare the number of United States Nobelists prior to World War II with those since.

From 1901, when the first awards were given, through 1944, there were forty-two chemistry prizes. Of this number three (7.1%) went to U.S. chemists; between 1945 and 1986, out of sixty-four chemistry laureates, twenty-seven (42.2%) were Americans.

In physics, during the first forty-four years, forty-eight awards were made and eight (16.7%) went to U.S. physicists; three of these were during the war years. For the years 1945–1986, out of a total of eighty-one prizes, forty-one of them (50.6%) were for work done in U.S. laboratories.

In physiology or medicine, the numbers are equally dramatic. Prior to 1945, out of forty-six awards, only nine (19.6%) went to the United States; for the period 1945–1986, Americans received fifty-four of the ninety-seven awards, or 55.7%.

Before the end of World War II, a policy decision was made by the

United States government to maintain research momentum and to emphasize fundamental research by continuing the partnership that had developed when the government began to fully appreciate its dependence upon the universities during the war. The decision reflected an appreciation that the partnership was effective not only for meeting military needs but to provide new information needed to wage war on disease and to fuel industrial machinery. It also recognized the necessity to find means of disseminating new scientific knowledge and to discover and train the best scientific talent available to ensure the future productivity of basic scientific research in the nation.

Thus, the cross-pollination that took place almost a hundred years ago in Europe between pure and applied science, between the pragmatic industrial sector concerned with a product and the academic world that cared only for "pure" knowledge, came into full bloom in the United States. The progeny and kin of this unlikely mating exist all over the world today, where a continuing flow of new technology and instrumentation affects the way the search for new knowledge is carried out. And these laboratories, like individuals of mixed ancestry who never fully resolve the conflict between a spiritual tie with one land and the practical necessity that binds them to another, are complicated organizations with a dual nature.

THE PARADOX OF SCIENTIFIC RESEARCH

Failure to recognize this duality and to take it into account can be the downfall of the scientific administrator. The scholar who refuses to recognize the exigencies that impose good business practices on laboratory management, or the manager who thinks he or she must turn the lab into a production plant, are equally doomed to failure. Individual dreams and scientific aspirations may recognize no boundaries, but plans for a real trip into uncharted areas must take into account fiscal reality, social responsibility, and political sanctions.

This scientific catch-22 was aptly and a bit cynically expressed by the first astronaut to venture into outer space. When asked upon returning what his last thoughts were just before blast-off into the unknown, he replied, "I remembered that every item upon which my survival and the success of the mission depended had been provided by the lowest bidder."

Ultimately, each scientist must deal with this paradox in his own heart and in his own way. But the responsibility for creating an institution that recognizes and resolves the dilemma—at least to the extent that it can be resolved—rests squarely on the shoulders of the institutional head, the chief administrator.

Whatever title he or she may hold, it is that individual who is responsible for creating the environment in which researchers function. It is that person who must balance the concept of "science," with its implication of venturing into the unexplored realms of knowledge and unlocking the secrets of the physical world, and the necessities of the organizational and bureaucratic world in which modern institutions exist. The style with which this is done and the extent to which it is successful will determine whether the research laboratory flourishes or founders.

REFERENCES

1. Crowther, J.G. *The Cavendish Laboratory 1874–1974.* New York: Science History Publications, 1974. p. 75.
2. Bush, V. *Endless Horizons.* New York: Arno Press, 1975. p. 172–173.
3. Ibid. p. 71–72.

Chapter 2

SCIENTIFIC ADMINISTRATORS: WHO THEY ARE AND WHAT THEY DO

To endure is greater than to dare; to tire out hostile fortune;
to be daunted by no difficulty; to keep heart when all have lost it;
to go through intrigue spotless;
to forego even ambition when the end is gained.

—William Makepeace Thackeray, *The Virginians*

Scientific administrators are of comparatively recent origin. When groups of scientists began to be collected under one roof to use common facilities, exchange ideas, and expose their results to their peers for discussion and approval, the research institution as we know it today came into existence. Before that, an eminent scientist might surround himself with students, technicians, builders, or other helpers, but they were all working on his ideas and functioned under his domination. In some cases, scientists worked entirely alone, providing all the skills necessary to their projects.

Every working group, however small, in which facilities must be shared, funds allocated among the constituent units, and individuals are dependent upon the knowledge and skills of one another, must have a leader, who may be called a director, head, manager—or less elevated characterizations from time to time.*

THE SELECTION PROCESS

There is no formula for selecting a scientific administrator and no training for managing a research institution except in the doing of it. The role, like

*For purposes of clarity, the titles used here to indicate the head of a scientific laboratory will be: laboratory head or director, and scientific administrator—the person who has overall responsibility for the achievements of the institution. When the titles research services manager or administrative manager are used, they indicate the staff member with responsibility for providing all services to the research staff—administrative, technical, engineering, maintenance, and so forth.

the title, develops uniquely within every organization, and no two are exactly alike.

A high-level member of the United Nations Secretariat, when called upon to describe the job of the Secretary-General, replied, "There is no such thing as the 'Secretary-Generalship' for which a job description can be given. It is something that develops as a result of the interplay between the person and the circumstances that exist at the time." So it is with leaders of research institutions. No two laboratories are alike, and the same laboratory changes so much over time that the ideal manager at one point in the life of the institution might be ineffective, or worse, at another stage.

The processes that bring about the selection of a laboratory head are somewhat like those that lead to the change of leadership in a sovereign state; new regimes are established either by peaceable process or by violence. The peaceful process in a research institution most commonly occurs when a leader moves on to another position. Occasionally, there is such deep dissatisfaction with leadership that insurrection brings about the leader's downfall and another ascends who represents the "new regime." The new regime establishes itself and, in a surprisingly short while, becomes the "old regime," exhibiting some of the same flaws and resistance to change as the old "old regime." Unless something occurs to stabilize the institution, another wave of violence will erupt and the whole drama will be played out again.

While anything that interrupts the flow of routine is upsetting, change is essential in scientific institutions. The advance of scientific knowledge in the past four or five decades has been like an avalanche sweeping over advanced nations and affecting the entire world. Research programs or institutions have been compelled to keep abreast of very rapid developments and to move forward with new discoveries or find themselves left behind. Hence, institution leaders have had to become masters in the art of managing change in a way that ensures progress. Alfred North Whitehead said that the art of progress is "to preserve order amid change and to preserve change amid order." Stability and routine are essential for good research but every nugget of new knowledge that comes to light has the potential to point the seeker in a new direction.

The alert and skillful laboratory director must be able to sense when to move in a new direction, while being aware of the dangers of being bedazzled by every exciting new idea or theory that catches the fancy of the scientific community or the public. And when the time comes to move toward a new direction, which often means phasing out or reducing support for one line of research in favor of another, the laboratory director needs the tact of a diplomat, the charm of a siren, and the conviction of the convert. Anyone can increase the budget of a research program, add to the staff, and offer more lab space; it takes genius to reduce all these things and retain the loyalty of the staff.

NOTABLE LABORATORY DIRECTORS

Successful directors have come from widely differing backgrounds and experience. Some Nobel laureates have succeeded while others have failed when given administrative posts. Yet, surprisingly, undistinguished researchers have become outstanding administrators who provide the setting for talented researchers to produce brilliant work. Autocratic, tyrannical laboratory heads have driven their colleagues and assistants to glorious achievements, and kindly, paternal leaders have also encouraged their groups to the same pinnacles.

Thomas Alva Edison, with only three years of formal education, employed more than seventy workers in his Menlo Park, New Jersey, laboratory. He was 24 years old when he set up his first lab and assembled about fifty men, mostly older, to work for him; still they called him the "old man." Edison's biographer Matthew Josephson says Edison once locked the doors to keep his workmen inside until they completed the job on hand—a rush job because he was trying to beat out a competitive inventor—thus turning the lab into a prison. Some of their wives came to the barred doors, wailing, or tried to convey parcels of food to the men through the windows, but Edison did not relent until the problem was solved and the job finished. When asked about some of those excessively zealous early days, he is quoted as saying, "At least I did not have ennui."[1]

The Edison General Electric Company became one of the parent companies that formed the General Electric Company in 1892, but Edison's leadership style apparently was not passed on to his successors. Willis R. Whitney, who directed research at GE for three decades, would go through the laboratory saying, "Are you having any fun?" Irving Langmuir (1932 Nobel laureate in chemistry) once replied to him, with great seriousness, "Yes, but what am I doing for GE?" Whitney replied, "That is not your worry!"

Some research administrators find that they cannot continue their own research and manage a laboratory effectively; others seem to be able to do both, although to do so requires abundant energy and limitless dedication. John R. Vane, who became director of research and development at Burroughs-Wellcome in 1973 after eighteen years of productive research at the Royal College of Surgeons in London, continued with his own active research program. When asked how he managed it, he said that he never took vacations, seldom left his office for lunch, and met with his research staff at the end of the administrative day, 6:30 P.M.. To the question, "But doesn't that make a very long day for your research staff?"* he replied matter-of-factly, "Yes, of course." In 1982, Vane was awarded the

---

*Statements by contemporary scientific figures that are not footnoted in the text were obtained in conversation with the author.

Nobel Prize in Physiology or Medicine, jointly with Sune Bergström and Bengt Samuelsson, both of the Karolinska Institute.

James Clerk Maxwell, designer, launcher, and inspirer of Cambridge's Cavendish Laboratory, continued his research while overseeing construction of the Cavendish, accommodating to the circumstances and inconveniencing no one. Maxwell's treatise, *Electricity and Magnetism*, which appeared in 1873, has been hailed as "one of the most splendid monuments ever raised by the genius of a single individual." While he was engaged in that work, he wrote in a personal letter, "Laboratory rising, I hear, but I have no place to erect my chair, but move about like the cuckoo, depositing my notions in the Chemical Lecture-room 1st term; in the Botanical in Lent, and in Comparative Anatomy in Easter."[2]

In contrast, Sir Brian Pippard, in speaking of his transition from research scientist and professor to head of the Cavendish Laboratory, said, "I recognize that I am now a full-time head of a laboratory, and I enjoy it. This is another phase of my career in which I work toward encouraging others and providing the atmosphere and resources for a number of scientific activities to go on. It is a different achievement from individual and independent research, but very interesting and rewarding."

One research administrator suggests that the

> breaking point is somewhere between 100 and 200, after which the scientist can no longer be one of the research staff who has been appointed leader, but must become an administrator. After as much as 1,000 or 2,000, the manager is an executive who cannot deal with the individuals, but must concentrate on appointing the right people in the next echelon.

From time to time, the question arises as to whether it is essential for the head of a research institution to be a scientist, since much of the job involves management and requires at least a glimmering of good business practices. It is not unusual in England for people with no science training to be appointed to administrative posts in scientific organizations. In the United States, this is exceedingly rare, and administrative posts in laboratories—fiscal, personnel, publications, and others—may be held by technical people. Robert Wilson, the first director of the Fermi National Accelerator Laboratory at Batavia, Illinois, placed physicists in nearly all managerial posts—business manager, personnel manager, and the like. He instituted a system of rotation, moving managers from one job to another and, in some cases, back to the research bench. He reasoned that such a system prevented isolation and empire-building while encouraging loyalty to the entire laboratory and not just one unit.

There is a school of thought that believes scientists do not like being administrators and that they have no training or aptitude for it, and it has even been suggested that saddling scientists with onerous administrative roles is a kind of punishment for having achieved scientific prominence. As one expressed it, "There is a saying that if one achieves eminence through outstanding research, a conspiracy develops to ensure that he never does it again." The first move of the conspiracy, according to this observer, is to persuade successful scientists to sit on committees, preside over conferences, hold office in societies, and (the coup de grace) head up research organizations, which is sure to keep them out of the laboratory where they might stand a chance of doing more outstanding work. The assumption is that since scientists have no training or talent for administration, they will botch the job. The assumption is wrong; many scientists who became noted for their research have gone on to become outstandingly successful administrators. The brilliant leadership of James D. Watson of the Cold Spring Harbor Laboratory on Long Island, where he has been director since 1968, comes to mind. (Watson, together with Francis Crick and Maurice Wilkins, was named a Nobel laureate in 1962 for the discovery of the molecular structure of DNA, the "double helix.")

If one accepts at face value the self-portrait that emerges from Watson's story of the days at the Cavendish Laboratory where the DNA research was done, Watson is about the last person one would expect to find managing a research institution or accepting administrative responsibility at any level. In his book, *The Double Helix*, Watson seems to be a young, irreverent, unmanageable genius who preferred going to dinner parties and the cinema, dating *au pair* girls and other young Cambridge women, and spending long afternoons talking with his colleagues over countless cups of tea (or stronger beverages) to spending dreary days toiling away in the laboratory.

Another great administrator, Victor Weisskopf, born in Vienna in 1908, made a name for himself as a young man with his important work on problems of radiation in quantum theory, and later became director-general of CERN (the European Center for Nuclear Research near Geneva) and chairman of the large physics department at MIT. In *Adventures of a Mathematician*, S. M. Ulam wrote that it was Viennese insouciance, combined with the highest intelligence, that enabled Weisskopf to navigate through not only the usual difficulties of administrative and academic affairs, but also in the more abstract realm of intellectual and scientific difficulties of theoretical physics. Ulam explains "insouciance" this way: "In post–World War I Berlin, people used to say, 'The situation is desperate but not hopeless'; in Vienna, they say, 'The situation is hopeless, but not serious.' "[3]

The large, federally funded, industrially operated Oak Ridge National Laboratory was headed for almost 20 years, from the middle 1950s to the middle 1970s, by Alvin Weinberg, who has been named to both the National Academy of Engineering and the National Academy of Sciences for his research achievements. Weinberg is often called a "statesman of science" because of his dedication and efforts toward the application of science and technology to the service of mankind. In 1984, speaking as a member of the "Management for Creativity" panel at the "Creativity in Science" conference at Los Alamos National Laboratory, Weinberg pointed out the specific talents of three scientific managers whom he described as "extremely successful":

> I think of Lee Hayworth [sic], the late director of Brookhaven; I. I. Rabi once said that Lee's very great power and strength was his self-lessness, that he was able to submerge himself for the good of the laboratory, which I think is a very important attribute of good managers. . . . Another person that comes to my mind is Eugene Wigner, who was the research director at Oak Ridge National Laboratory for a year and a half. In Wigner's case, his enormous intellectual power was contagious and the people around him were inspired by him and wanted to emulate him. Now, usually, they couldn't emulate him but in trying to emulate him they worked at the edge of their competence.
>
> The third person who comes to my mind is Alex Hollaender. For many years, Alex was the director of the largest biology division in the Atomic Energy Commission. It was the most productive biology division, at least measured by the number of National Academy of Science members who came from there—no fewer than sixteen. Alex's great secret was that he *knew* his people were the best in the world. He knew it and he conveyed that sense to them, especially to the young people. . . . Since they were obviously the best in the world, Alex would insist that the people in his division had to work harder than anybody else, and in fact, the biology division did work harder than the other divisions of Oak Ridge National Laboratory.[4]

George W. Beadle, a much admired and loved scientific administrator, was a postdoctoral fellow in the Division of Biology at California Institute of Technology from 1931 to 1933, and thirteen years later he returned as professor and chairman of the Division of Biology. From 1961 to 1968 he was president of the University of Chicago. Beadle and Edward L. Tatum of Rockefeller University were awarded the Nobel Prize in Physiology or Medicine in 1958 "for their discovery that genes act by regulating definite chemical events." Beadle often replied to official correspondence in longhand messages on small notepaper and kept no copy. It delighted his correspondents to receive such prompt and informal response from such a busy and noted person, but his office staff had to be very alert to keep track of engagements he accepted in that manner. He had no patience with formality and bureaucratic time-wasting and fulfilled his heavy administrative responsibilities for many years while continuing to carry on an active research program.

Willard F. Libby held numerous distinguished administrative posts in both industrial and academic organizations. He was awarded the 1960 Nobel Prize in Chemistry for developing a method of radiocarbon dating and received many other scientific awards and honors. He was director of the Institute of Geophysics and Planetary Physics at the University of Chicago and directed research at Douglas Aircraft Company, among other institutions. He also served as a member of the advisory board of the Rand Corporation.

There have been few notable women scientific administrators. Until quite recently, women were either openly or covertly discouraged from pursuing scientific careers. Today women scientists head up scientific societies and important committees and are beginning to be given significant administrative appointments.

One of the earliest and outstanding scientific administrators among women was Mina Rees, a mathematician, who held a key post in the Office of Scientific Research and Development during World War II, for which she later received the King's Medal for Service in the Cause of Freedom from King George VI of England, and the President's Certificate of Merit from President Truman. Rees headed the mathematics branch of the Office of Naval Research from 1946 to 1949, when she was appointed director of ONR's mathematical science division. She became the first president of the Graduate School and University Center of the City University of New York in 1971.

Dixy Lee Ray, a marine biologist who was named to the Atomic Energy Commission in 1972 and in 1973 became head of the commission (the title was "chairman"), was as much a politician as a scientist. After leaving the AEC in 1975, she served as Assistant Secretary of State and later was elected Governor of the State of Washington.

Early in 1987, the Carnegie Institution of Washington appointed Maxine Frank Singer as its eighth president beginning in 1988. A molecular biologist who is currently chief of the biochemistry laboratory at the National Cancer Institute, Singer is widely recognized for both scientific and administrative achievement. She is a member of the National Academy of Sciences, the American Academy of Arts and Sciences, and the Pontifical Academy of Sciences at the Vatican.

When Professor (Sir Nevill) Mott, head of the Cavendish Laboratory at Cambridge from 1954 to 1971 was asked to comment on scientists as administrators, he replied, "I think every scientist should be able to function in the three-part role of professor, researcher, and administrator. I found it interesting and exciting. I certainly do not agree that scientists have no interest in or talent for administration."

Comparisons can be made between those who rise to leadership roles in scientific institutions and those who reach the top executive ranks in corporations. Styles change in executives as society's so-called psychostructure brings certain individuals to the forefront and relegates others to the obscurity of midlevel management, while the majority are led, manipulated, handled, managed, motivated, structured, and organized into units where cooperation and compromise are rewarded with approval— or at least the absence of disapproval—and a sense of something referred to as "security," which, of course, does not exist.

Scientific and industrial leaders alike are often marked by intelligence, courage, industry, and the ability to gain the confidence of others. One major difference is that industrial tycoons usually have no reticence about admitting their struggles to reach the top, whereas even those scientists who wind up heading the most prestigious institutions often disclaim any early ambitions to do so; they seem to take pride in implying that, as has been said of greatness, administration was "thrust upon them."

Although few will admit to it, no doubt there are some scientists who early in their careers aspire to heading up a great institution. C. P. Snow's fictional physicist, Arthur Miles, mused after the presentation of his first paper before the Royal Society:

> I liked the affability of the High Table, and the wine. I caught myself wanting to bask in it all; in self-defence, ashamedly, I kept dwelling on the ambition that was a step nearer, now. I thought of running a scientific institute, how I would begin, get my men, help them work. . . .[5]

Snow, himself a trained scientist, knew scientists and the scientific world well enough that it is unlikely the character of Arthur Miles was created out of whole cloth.

SCIENTISTS VS. NONSCIENTISTS AS LABORATORY DIRECTORS

There are powerful arguments favoring the choice of a scientifically trained laboratory head. No research institution is an island; every laboratory must keep abreast of what is going on in other places doing similar or relevant research, and in order to draw effectively from the broad world of science one must be a part of science; an established competence in some area of science is the "ticket of admission" to that world. Moreover, the head of an institution is its external representative, and the image that person presents affects the way the outside world views the organization. The statements, decisions, and attitudes of the leader are presumed to be those of the laboratory.

The specific areas in which scientific training is an advantage to the head of a laboratory are:

1. Interacting with other scientific institutions;
2. Making key decisions about research programs;
3. Understanding safety issues—personnel protection, waste management, and so on;
4. Communicating with science writers;
5. Comprehending the significance of critical facilities—instruments, libraries, and so forth;
6. Continuing education—organizing courses, keeping up with seminars and workshops at other places, arranging exchanges.

A very small number of outstanding administrators of research institutions have not been scientists, but they are few and far between. Joseph Slater, who became president of the Salk Institute in 1967, was not scientifically trained. His previous career had included administrative posts in the State Department, the Ford Foundation, and the U.S. Occupation of Germany; but in a very short time he infused the laboratory with an entirely new spirit that, in retrospect, seems to have turned it in the right direction. The institute was created to honor Salk and was heavily funded by the National Foundation, formerly the Foundation for Poliomyelitis, that supported Salk's vaccine work. It was unthinkable that anyone except Jonas Salk would head it up, but he wanted to continue his research activity, and so he defined his role according to his own interests and capabilities. He spent the first few years using his persuasive charm and the inducement of a beautiful, modern building to attract some of the best scientists in the world to La Jolla. Recognizing that it would be impossible to act as administrator and fund-raiser (the Salk Institute has no endowment and no public support) and actively recruit the scientific staff while continuing research, Salk persuaded the institute's governing board to appoint Joe Slater as president and give him responsibility for general management, particularly for bringing about and maintaining the financial stability of the institution. This organizational arrangement continued until 1984 when Salk retired. He still retains the title of Founding Director.

The two-headed management plan worked well for the institute, judging in the only way one can judge a scientific organization, that is, by its scientific output. Joe Slater had to overcome formidable financial difficulties in order to bring Roger Guillemin and his research group to the institute in 1970, but his conviction that it had to happen, along with his irrepressible self-confidence, brought it off, and it proved to be worth the effort. Guillemin and Andrew Schally of the Veterans Administration Hospital (New Orleans) were awarded the Nobel Prize in Physiology or Medicine in 1977 for their discoveries concerning the peptide hormone production of the brain. One of the first scientists Salk brought to the lab, Renato Dulbecco, was named a Nobel laureate in 1975. Robert W. Holley, who went to the Salk Institute from Cornell in 1968, was jointly awarded the Physiology or Medicine Nobel Prize that same year with Har Gobind Khorana of the University of Wisconsin and Marshall W. Nirenberg of the National Institutes of Health, Bethesda, Maryland.

QUALIFICATIONS

The best ploy for lifting morale and for coalescing disgruntled groups is to create a feeling of excitement, a sense that new things are happening or just about to happen. Alexander Hollaender at Oak Ridge was a master of this, as was Slater at the Salk Institute. They always reported the good news first and emphasized it; then, if there was bad news to report, they managed to make it seem relatively insignificant and nothing their superb research organizations could not handle.

Institutional heads get their jobs in one of three ways: in some cases, they are promoted into them, after ascending the scale of administrative positions. In others, they are plucked out of the laboratory with no experience in management beyond their own research groups. The third way is by recruitment conducted by "search committees." The first thing a search committee does is write up a job description and a list of qualifications. But, like the position of the United Nations Secretary-General, there is no such thing as a job description to describe accurately the functions of a scientific administrator. So much depends upon the situation of the institution at the time and the interaction between its constituent elements and the person chosen. The best that can be done is to draw up a list of responsibilities—direction of research program, maintaining financial stability, and formulating or continuing an educational curriculum. Minimum qualifications always include an educational requirement, usually a graduate degree in a specific field. Scientific standing and reputation cannot be listed as qualifications, but they might as well be because they certainly count. A fairly extensive, or at least an acceptable, record of publications may be on the list. These basic requirements are worth little; it is quite possible for a candidate to meet fully the educational, publication, and public recognition requirements and still not be able to manage a laboratory. The only use such a list serves is to screen out those without the barely minimum qualifications.

An examination of credentials requires evaluative judgment, particularly of the candidate's personality in relation to the institution, and it helps to have faith, ESP, a sixth sense, and more black magic than most people can conjure. Is a Ph.D. in physics better than one in chemistry, biology, or engineering for this post? Are five publications each in journals X and Y better than ten in Z? Will this person be able to inspire the research staff or drive them to work at the peak of their creativity? Or, will this candidate have the courage to risk supporting a new idea and the strength and diplomacy to say no to a bad one without destroying the morale of the researcher involved? Will he or she be able to discern the good idea from the poor one? These are not easy questions.

The Laboratory Director's Responsibility

Each laboratory has its fundamental purpose and goals from which specific administrative functions emanate, and no two will ever have the same needs in leadership. But, generally speaking, the duties of a scientific administrator fall into the following categories:

1. To propagate a high level of research or technological development. To see that the preestablished objectives of the organization are fulfilled. To constantly examine those objectives and provide leadership neces-

sary to alter, abolish, or reroute them as new scientific information comes to light.

2. To assemble the fiscal resources necessary to carry out the objectives and to maintain the financial stability of the institution.

3. To recruit a highly qualified research and administrative staff.

4. To provide and maintain a working environment that will evoke from every staff member his or her highest level of performance. Such an environment includes the provision of laboratory space, administrative services, the necessary facilities and equipment and their maintenance; communication facilities for disseminating research results and learning about scientific achievements in other laboratories around the world; a safety program appropriate to the hazards entailed in the work.

5. To establish and maintain a system of internal communication, and to promote the free exchange of information among the members of the research staff, particularly among those whose work overlaps the work of others. At the same time, to protect the security of work going on in laboratories where such protection is desirable and justified.

6. To provide for dissemination of information to the scientific community and the public. The achievements of the research staff will be recognized by other scientists only in so far as the results of that work are disseminated. The public must be informed and educated about the purpose of the laboratory, with appropriate safeguards for the security of the work, particularly if there is any possibility that some activities might be conceived as having a deleterious effect within the community. An informed public is very important in a society in which the public provides most of the research money.

7. To establish and maintain the ethical and moral health of the institution.

8. To serve as the laboratory external representative and to speak for it.

There may be some variation in the emphasis placed on these functions and the priority given to each within different institutions, but the propagation of a high level of scientific work in the pursuit of the objectives of the institution always comes first.

DECISION-MAKING

The successful fulfillment of the responsibilities implied in these categories requires broad knowledge and command of many skills, but the key quality is the ability to make wise decisions. Epictetus is credited with this description of a philosopher: "a will undisappointed; evils avoided; powers daily exercised; careful resolutions; *unerring decisions*" (italics added). "Philosopher" in Epictetus's time included scientists, and indeed this might well describe the ideal director of a research institution, although few, if any, will always be able to avoid evils and hand down unerring decisions.

The talent for making good decisions is largely innate. In *Modern Man in Search of a Soul*, Carl Gustav Jung wrote, "The great decisions of human life have as a rule far more to do with the instincts and other mysterious unconscious factors than with conscious will and well-meaning reasonableness. . . ."[6] Instinctive actions taken in opposition to all known data turn out to be brilliantly accurate often enough to prove that there is wisdom that transcends logic and analysis. Such was the wisdom demonstrated by Adriano Buzzati-Traverso when the International Laboratory of Genetics and Biophysics moved to its new buildings in Naples, where assignment of parking spaces became a very acrimonious issue. After considering all the possible systems, Buzzati came up with an inspired solution: spaces were to be assigned on the basis of age—the older scientists would be given those closest to the laboratory. He then sat back and enjoyed observing the pride with which younger staff members walked to the farthest points in the parking lot and announced their parking assignments often and loudly. But the Solomonic instinct is not given to all, and those who lack an inner guiding light must learn to mobilize guidelines or other assistance in order to arrive at wise decisions, or at least avoid the disastrous ones.

Decision-making requires, above all, the ability to perceive accurately which matters are seminal and which can be dispensed with summarily. Some will be obvious but, occasionally, a seemingly trivial decision can reverberate through the institution with shocking impact. Timing can be critical; a matter that may be minor at one time can turn out to have widespread effect at another. Institutional heads must learn to be acutely sensitive to the climate within and without their organizations.

Major issues sometimes appear in disguise, and over a period of time, the nature, quality, or focus of a research program may be transformed by hasty or careless decisions concerning matters that seemed relatively unimportant. The quality of research emanating from a laboratory does not reach excellence overnight, nor does it deteriorate in that space of time. But, slowly, a series of unwise and seemingly minor decisions may pile one upon the other and by the time the laboratory itself and the scientific community around it realize what has happened, it is often too late. As King Solomon said, it is "the foxes, the little foxes, that spoil the vines."

Major decisions for the scientific administrator fall into one or more of these categories:

1. The objectives of the laboratory; any significant change in the research program's direction or level of funding; introduction of a new line of research or termination of an ongoing program.
2. Matters that affect the whole institution, scientifically and administratively.
3. Matters that go beyond the laboratory in their effect, reaching into the

surrounding community, the wider scientific community, or other branches of the institution with which the laboratory is affiliated.

4. Matters that set an important precedent. Precedent-setting is the way leaders put their own stamp upon an organization, but drastic departures from past practices should be given the most careful study and consideration.

5. Any matter that could seriously affect the morale of the group.

Some administrators seem to be gifted with instinct or that "mysterious unconscious factor" Jung referred to, which, in addition to their superior intelligence and broad knowledge, qualifies them to become superb leaders even when they themselves do not fully understand how they do it. And it is not unusual for highly regarded scientific administrators to disclaim any responsibility for the success of what occurred in their laboratories. This quality is beautifully illustrated by a conversation with professor J. S. Mitchell, who was for many years director of the Radiotherapeutics Center at the University of Cambridge. When asked to comment on the functions of the scientific administrator, his offhand reply was "I don't have any answers; you just do it." When pressed further for specifics, the conversation went like this:

Q: Why do some scientists build superb institutions, in the face of great odds, and others take over a good one and destroy it? What professional, personal, or other characteristics make the difference?

A: I don't have any answers.

Q: Did you have good subordinates to whom you could delegate?

A: Oh, no. I had terrible subordinates.

Q: And in your lab—did you not have a good lab chief?

A: Oh, yes. I have always had a top-notch lab chief—couldn't do a thing without one.

Q: Would you say that is one factor?

A: Well, yes, I suppose so.

Q: Do you have to present an annual budget?

A: Yes, sure.

Q: And how do you put it together? Do you ask the various lab heads, senior scientists, and so forth to present a budget?

A: Yes, and we talk all the time. We discuss everything.

Q: How often do you meet?

A: No regular meetings. We see each other in the hall; I go to their labs.

Q: You go to their labs?

A: Oh, yes, it's the German tradition. You speak to everyone in the lab every day.

Q: Everyone? Every day? Dishwashers, animal handlers, technicians, janitors?

A: Yes, of course.

Q: How do you recruit new people?

A: We advertise.

Q: And do you get applications from good people that way?

A: Yes. Our top physicist came to us that way.

Q: Don't you use the old-boy network?

A: Not at all.

Q: You never call up your friends and tell them you have an opening and ask if they have anyone to recommend?

A: Oh, yes. I do that.

Q: What proportion of your appointments are made through advertising?

A: Well, a lot of our people are trained here and we fill our places from within.

Q: What proportion from within and without?

A: Fifty-fifty.

It was like asking a high-wire performer how he does it, who replies, "It's nothing. You just start from this side and walk over to that side."

REFERENCES

1. Josephson, M. *Edison: A Biography*. New York: McGraw-Hill Book Co., 1959. p. 90.
2. Crowther, J.G. *The Cavendish Laboratory 1874–1974*. New York: Science History Publications, 1974. p. 48.
3. Ulam, S.M. *Adventures of a Mathematician*. New York: Charles Scribner's Sons, 1976. p. 261.
4. Raju, M.R., Phillips, J.A., and Harlow, F. (eds.) *Creativity in Science—A Symposium* (LA–10490–C). Los Alamos, NM: Los Alamos National Laboratory, 1985. p. 102.
5. Snow, C.P. *The Search*. New York: Charles Scribner's Sons, 1958. p. 108.
6. Jung, C.G. *Modern Man in Search of a Soul* (trans. W.S. Dell and C.F. Baynes). New York: Harcourt Brace and Co., 1933. p. 69.

Chapter 3

THE ADMINISTRATIVE STAFF

Genius . . . means transcendent capacity of taking trouble.
—Thomas Carlyle, *Life of Frederick the Great*

In some laboratories administration is viewed as a hindrance—not a help. It's something to be got around, bypassed. At the "Creativity in Science" symposium, held in Los Alamos in 1984, Stewart Blake observed, "I know of cases where the people in administration, responsible for efficient management, tend to look upon the laboratory as existing to support the administrative function rather than the administrative function existing to support the work of the laboratory. . . . One of the great responsibilities of senior and middle management in laboratories is to control the administrative function so that it does as little violence as possible to the degree of independence needed by researchers."[1]

The responsibility for controlling what scientists sometimes refer to as "what our management is doing to us," rests squarely on the shoulders of the laboratory head who establishes the policies for the administrative management of the institution and appoints the staff to carry out those policies.

RESEARCH SERVICES

One of the most important appointments the head of a research institution makes is the administrative manager; someone who directs the nonscientific functions—accounting, personnel, purchasing, publications, and plant engineering (facilities and general maintenance)—in a manner that enhances the productivity of the research staff. In a large organization, it may be a director of administration who selects departmental heads for the various functions and oversees their operations. In a small laboratory, a business manager or administrative aide may handle one or more functions, such as accounting and personnel, and supervise the work of a few subordinates who take care of other activities.

"Someone to handle the paperwork" is a frequently heard description of the administrative services chief, and the laboratory director who considers that ability the sole qualification for the post is probably going to hire someone who can do just that and nothing more. It is understandable—even if unforgivable—that this happens. The laboratory director is so deeply involved with the primary task of assembling a scientific staff, particularly at the outset, that it is easy to relegate nonscientific appointments to the back burner. Few directors are visionary enough to see the wisdom of taking care, even under the pressure of organizing a new laboratory, to consider the contribution that can be made by nonscientific staff to provide an ideal research environment. The professional administrator in a research laboratory fully comprehends that everything done by the nonscientific staff aims to reduce the demands made on the scientists, even if it means, as may well be the case, that the administrative staff assumes a heavier burden.

When he was director of research services at North American Aviation Science Center (now Rockwell International Science Center), Jack Balderston defined the role of the administrative staff as one of helping the research staff accomplish *their* jobs as expeditiously as possible. This might be described as "service-oriented management" as opposed to "control." "The attitude of service," Balderston wrote, "should be, 'How can I help?' . . . Control is oriented as saying 'No.' Control means enforcement, and the rules, procedures, or policies are inviolate. The control-oriented administrator sees that procedures are followed to the letter and forms filled in completely since they are the means of enforcement. The service view is that when forms and procedures become nothing but a stumbling block, they should be bypassed or abolished."[2]

The professional administrative staff necessarily understands and accepts the dominant role of the scientific staff and must be content with a kind of second-hand satisfaction from the scientific accomplishments of the laboratory. To do this, they need to be aware of what the scientists have done and are doing. At Oak Ridge's biology division, the administrative staff was encouraged to attend seminars when they felt inclined to, and while much of the material was over their heads, those who chose to attend had a better sense of the work and were more likely to identify with it than those who did not. At the North American Aviation Science Center in Thousand Oaks, California, Jack Balderston initiated a monthly series of lectures at which scientists would discuss, in lay language, the work they were doing. The entire administrative staff was invited, and many attended. The lectures were followed by a tour of the speakers' labs. To reverse the process, the administrative staff prepared a monthly newsletter to circulate among the scientific staff. It contained items about the nontechnical side of the laboratory.[3]

The relationship between the head of a laboratory and the administrative professional is central to developing a mutually satisfying working rapport between scientific and administrative staffs. The administrator must be someone in whom the director can place complete confidence and who can also inspire the confidence of the rest of the staff. The personality of the individual chosen for the position is as important as administrative qualifications. The laboratory head and administrative chief must, above all, have a compatible philosophy about the management of a research institution. If the director is control-oriented and insists that the established procedures be followed to the letter with no deviation, he or she will find a service-oriented administrator unacceptable. The reverse situation can sometimes work. That is, if the administrative professional sees the role as that of the "enforcer" and thinks "No" is always the first answer, the laboratory head can become the "good guy" by countermanding the subordinate's decisions from time to time. The risk is that the scientific staff will lose respect for the administrator and that more and more decisions will be appealed until the laboratory director, in effect, becomes the administrative chief. When this occurs, it is impossible to hold a competent individual in the chief administrative post, and the entire nonscientific staff comes to reflect the weaknesses of the person willing to accept such a position.

Sometimes there is a very fine line between decisions that fall within the purview of the scientific director and the professional administrator. In the final analysis, it is the director of the institution who makes the determinations. The closer the working relationship between the two, the less likely it is that conflicts arise. They must meet frequently, on a schedule suitable for the size of the laboratory, the nature of its objectives, and the areas of responsibility assigned to the administrator. Except in rare cases, there should be no secrets between them, and before overruling any decision of the administrative professional, the laboratory director must hear his or her side of the story.

Sir Nevill Mott, in speaking of his years as head of the Cavendish Laboratory, commented, "The possibility of remaining afloat as a laboratory or department head depends very much on having an able administrative assistant or aide." It is only with such assistance that the head of the laboratory can hope to pay full attention to the research activity of the institution, and it is indispensable for one who continues to be directly involved in research.

The ideal relationship develops, usually, out of years of working together, but it is possible to sense from the beginning whether such compatibility can be achieved. A new institutional head usually inherits a subordinate responsible for the general administrative services and must make the decision about that position soon after arrival. The incumbent may be quite competent and have long experience at the laboratory but, for some reason, the director may not feel that they will be able to work well together. In such a case, it is better to make a change in the most diplomatic and appropriate way possible. Transfers can sometimes be arranged when the laboratory is a constituent unit of a larger organization.

One newly arrived director made what might be considered a Solomonic decision, after assessing the professional administrator he inherited. He found the man to be very able in handling the paperwork aspects of the position, but brusque and tactless in dealing with individual staff members, and he was therefore very unpopular with the scientific staff. In order to resolve the problem without losing the expertise of the employee, the director added a second assistant and assigned to that person all administrative functions requiring personal dealings with the staff. Since the laboratory could afford a second person, it turned out to be a wise resolution. It represents an approach exactly opposite to the "new broom" espoused by some who, when taking over the leadership of an organization, sweep out key people and bring in colleagues from former affiliations. It also eschews the cautious approach, favored by some who hesitate to disturb anything for fear of doing something wrong.

These people may not rock the boat, but neither do they make much progress. Writing in the July-August 1977 issue of *Science*, Victor Weisskopf cautioned against the status quo philosophy: "There is a distinction between order in living and dead nature. At the very end of everything, when the sun is extinguished, matter will be even more ordered than it is now, because all random heat motion will be frozen. But everything will be cold, dead, and unchanging. It is the temperature gradient between the hot sun and the colder earth that produces the living order, ever changing and developing, through reproduction and evolution."[4] The "living order" philosophy of management may be ideal for the research institution.

The administrative staff in a research laboratory cannot enhance creativity or inspire enthusiasm, but it can help to generate the ambiance that is so important to successful science. It can also hamper creative effort and dampen enthusiasm by introducing frustration—a situation that must be avoided at whatever cost in effort and understanding on the part of those who make their contributions to science in nonscientific positions. The major areas in which these contributions are made to a laboratory are in accounting, personnel, purchasing, communications, and plant engineering (facilities and general maintenance).

The organization chart of the administrative staff of a large multipurpose institution is shown on page 29. It gives a clear picture of the functions performed by the nonresearch staff and is useful as a guide for an organization of any size.

ACCOUNTING

This function is sometimes called "Budget and Accounting," since it includes assisting the laboratory head and group leaders in budget preparation and submission. It is the institutional head who makes all final decisions about budget submissions to higher authorities and allocates funds among the various research groups when a final budget has been approved. The role of the accountant (or accounting staff) is to assist those who prepare operating and capital-expenditure budget proposals and to provide each group leader and the head of the institution with periodic budgetary status reports. The accounting department also assists in preparing budget estimates included with grant proposals and it must be fully knowledgeable about all aspects of fiscal requirements in connection

ADMINISTRATIVE STAFF ORGANIZATIONAL CHART

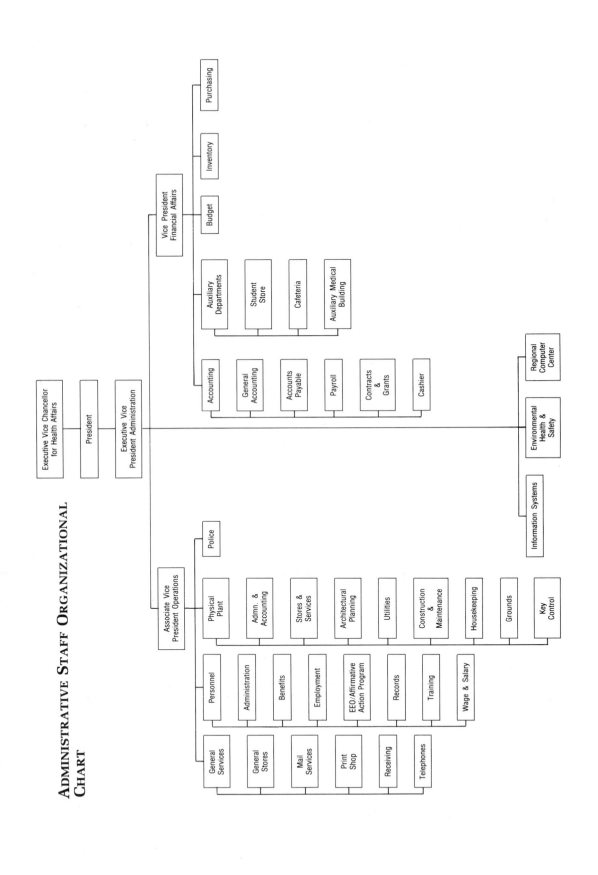

with outside sponsors. In laboratories where a great deal of research is done under outside sponsorship, there may be one or more accountants specializing in administration of grant funds and preparation of budget estimates to accompany grant proposals.

Before the accountant and the group leaders meet to discuss an operating budget, each must have certain facts or assumptions on which to base the submission. Requests for appreciable increases will have been ironed out by the laboratory director before the accountant is brought into the picture. The director may have agreed to "entertain a request for a ten percent increase over the previous budget period," or may have ordered a ten percent decrease, which provides a basis for the estimate. The scientist concerned makes the request, based on program needs and on any discussion with the director that may have occurred; that information is then reconciled with the input from the accountant.

The most creative scientists are often paralyzed at the prospect of making up a proposed operating budget (which usually must be presented about two years before the effective period). The results from a current experiment might alter the needs of the program for the next five years, and the reluctance to anticipate requirements six months in advance, much less two years, is understandable. Future overhead costs must be taken into account, and an estimate made on the basis of an anticipated rise or fall (if, indeed, there has ever been a fall). And every item is subject to that variable called "cost of doing business," which is nothing more than the effect of inflation or deflation. One does not have to be a prophet to arrive at an estimate; there are plenty of seers around, called economic forecasters, who are in the business of looking into the future and predicting what the inflation rates will be a year or two hence. The accountant can suggest a reasonable figure, and no matter how far off it may turn out to be, a firm working figure is reassuring to someone who is terrified at the thought of estimating costs of future operations.

The preparation of an operating-budget presentation is based mainly on these questions:

1. Will personnel be increased? Or decreased?
2. Will the requirements for supplies, noncapital equipment, travel, or general operating costs change appreciably?
3. How are overhead costs calculated? What portion is to be included for the program in question?
4. What is the "cost of doing business" (inflation/deflation) factor?
5. What is the justification for the continuation of the work at present levels or with an increase?

The most important part of a budget submission is the "justification," which must be written by the scientific staff or the scientist concerned. It is an updated progress report of what has been done and a forecast of what will be done in the fiscal period for which the budget is being prepared. A request for a sizable increase requires a much more elaborate justification than a submission that assumes a continuation of work at the same level of funding.

The justification should be composed with great care because, when the final budget is approved, it may well become the authorization for carrying on a particular line of research. Although it is seldom wise to mislead in these write-ups, this is the means whereby some research that may not be approved otherwise can be "bootlegged," that is, gain approval from higher authorities who are so strongly focused on a main goal, they overlook other peripheral objectives that may be simultaneously achieved. When the Atomic Energy Commission funded the biology work at the Oak Ridge National Laboratory immediately after the end of World War II, they were strongly motivated to find out as much as possible about the biological effects of radiation. Alexander Hollaender recognized that in order to study radiation effects reliably, a great deal had to be learned about basic biological, chemical, and physiological mechanisms in living things. While adhering to the AEC mandate to study biological effects of radiation, the budget justifications also included basic studies that were essential in order to assess the radiation effects fully. In this way, Hollaender built one of the best basic biology research laboratories in the world.

After the budget proposals have been completed by all research units or department heads, they can be combined into a budget for the entire lab and sent to the research services manager, who uses it as a basis for the preparation of the budget for administrative, nonscientific services. The research services manager also prepares a proposed capital-expenditures budget.

Capital expenditures include construction, alterations, large items of general-use equipment, and major scientific instruments, which, by the time they appear in a budget proposal, will have been thoroughly discussed, perhaps over a period of years. In addition, each department or group leader will include with the operating-budget submission large, costly items of equipment that are requested for use by a single program or group. Very expensive pieces of equipment, such as computers, ultracentrifuges, and electron microscopes, may also have been discussed with the director before they appear in the form of a budget request.

The one thing that can be said with confidence about the list of equipment requests is that the first one is always too long, more than the director is likely to approve for final submission. The director of research services will very likely discuss all the major items with those requesting them before submitting a tentative list to the laboratory director. In preparing the final submission, the laboratory director may call upon the research services director and the accounting staff for assistance, and upon members of the research staff for advice. There may be numerous *draft* budgets before a final one is agreed upon. However, if only those major changes that were discussed with the director prior to submission have been included, the bottom line will not be a surprise. It will be the final figure of the previous budgetary period, plus the anticipated increases in overhead and "cost of doing business," plus any specific increases or decreases authorized by the director.

As valuable as accounting assistance is in budget preparation, it may be second in importance to the compilation of *budget status reports*. Each department (or cost center) should receive a monthly status report indicating the expenditures to date, by category, and the balance available for the remainder of the year. A copy of each report and a one-page summary of them should be sent to the research services director and to the head of the laboratory. Computerization in accounting has made it possible to turn out voluminous reports with so much detail that they may discourage perusal; status reports should be summarized on one page, with additional pages to show breakdowns only if they are essential to the administration of the funds. They should indicate, at a glance, whether a unit is over or under budget and in which categories. The summary of these various reports will provide sufficient information to the director of research services and the head of the laboratory. At the halfway mark of the fiscal year, it will be obvious from these reports whether a budget revision is necessary, if austerity measures need to be instituted, or if additional funds must be acquired. In many organizations a midyear review is a standard procedure and the accounting department is expected to provide additional detail on the budget status for each cost center. It is quite possible that a group may be underspending in the first part of the year because they are awaiting deliveries or arrival of additional personnel. Midyear review information presents a more accurate picture than figures in monthly status reports. A sample budget status report form that was used successfully at one research laboratory is shown on page 33. It may not fit the needs of every organization, but it can serve as a guide to the design of a suitable one for your laboratory.

Unlike profit-making organizations, research laboratories are not oriented to the idea of the bottom line and their barometer of success is not profit; the product produced by a laboratory is new information, and good stewardship reflects a fair return of research results for the funds available. Inspired leadership often brings about a far greater than fair return. Careful administration of fiscal resources requires competent accounting assistance, without which even the most creative leader will have great difficulty realizing a maximum return in research results for available funds.

PERSONNEL SERVICES

The personnel department in a research laboratory is responsible for all the services under the title "employee relations": employment procedures, fringe benefits, medical and first-aid care, continuing education

Budget Status Report

TO: _____

Budget Status as of _____
ACCOUNT NAME _____ · ACCOUNT NUMBER _____
AGENCY NUMBER _____ · DEPARTMENT NUMBER _____
CURRENT PERIOD: FROM _____ TO _____

Ledger _____

COMMITMENTS & EXPENSE THRU DATE INDICATED →

| CATEGORIES OF COST | BUDGET | | | EXPENDITURES | | TOTAL TO DATE | | COMMITMENTS | | BUDGET BALANCES AVAILABLE |
	ORIGINAL	TRANSFERS	REALIGNED	CUMULATIVE TOTAL LAST MONTH	ACTIVITY THIS MONTH	Amt.	%	AMOUNT	THROUGH DATE	
SALARIES & WAGES										
SUPPLIES & EXPENSE										
EQUIPMENT & FACILITIES										
TRAVEL										
STAFF BENEFITS										
OTHER EXPENSE										
UNALLOCATED										
TOTAL DIRECT										
OVERHEAD @ RATE OF										
TOTAL										

Prepared by _____ Date _____

for nonscientific personnel, provision of recreation and cafeteria facilities, and any other related functions. The personnel manager may also fill the role of unofficial (or official) ombudsman. Should a conflict emerge between a member of the scientific staff and someone on the administrative staff, or between a scientist and a technician, often the personnel office is the first place it becomes known. The scientist or other person involved may come directly to the personnel office to discuss the matter or file a complaint. But even if this does not happen, frequent absences, visits to the dispensary for unspecified illness, tardiness, a poor work record, or a request for transfer are clues that often surface first in the personnel office. Not all scientists are human-relations oriented, and their absorption in their work sometimes blinds them to the attitudes and prerogatives of those who assist them. A sensitive and understanding ombudsman with a bit of talent for diplomacy may help bring about a rapprochement. Or, if things have gone too far, the problem may be resolved by personnel transfers or exchanges, a function that the personnel office should manage skillfully.

The personnel manager assists in staff recruiting, assembles necessary dossiers on candidates for positions, and arranges interview visits, but final personnel selections are made by the groups or departments where the staff members are to be placed. A very competent individual who has gained the respect of the staff for good judgment in evaluating the qualifications of candidates may participate in the decision. (Recruiting and appointments of the scientific staff are discussed in detail in chapter 4, "Research Staff.") The personnel requisition form, shown on pages 36–37, is used for both scientific and nonscientific personnel in a large research and teaching institution.

The personnel manager is appointed by the director of administrative services, who also gives final approval of all clerical, secretarial, crafts, maintenance, and animal-caretaking personnel appointments, after their acceptance by their supervisor. Occasionally, however, the laboratory head will have final approval.

The personnel manager must be well acquainted with the educational and recreational resources provided by the community and will be one of the most active members of the laboratory family in interacting with community organizations such as the local Red Cross chapter, YMCA, YWCA, high schools, and junior colleges (which often have educational courses and recreational facilities that may be open to the staff of the research laboratory). The personnel manager should explore these educational facilities and make use of them if practicable. The personnel office will also be in communication with various charitable funds and organizations, such as the United Fund and other annual fund-raising campaigns. Certainly, the personnel manager needs to be familiar with all local facilities such as fire department, police department, hospitals, and ambulance services, as well as personally acquainted with medical specialists who may be needed in emergencies. The personnel office might be thought of as the human-services department of the laboratory, and employees should feel that is the best place they can go for help with problems that do not seem to fall within the function of any other department.

The personnel manager must keep abreast of current legislation concerning such things as labor relations, affirmative action, fair labor standards, and policies regarding withholding from wages and salaries for tax purposes, garnishments, or charitable deductions. In large institutions, there will be a lawyer or a legal department to which questions can be referred, but the personnel manager must keep up-to-date on policies of the institution as well as state, local, and federal government regulations. Information on labor relations is available through the publications of the Bureau of National Affairs (BNA). Since 1937, BNA has published Labor Relations Reporter, recognized as a leading labor-relations information service. Information on BNA publications can be requested from: Bureau of National Affairs, 1231 25th St. NW, Washington, DC 20037.

Prentice-Hall issues a variety of publications in the personnel-management field: *Labor Relations Guide* and *Personnel Management: Policies and Practices* are two that the personnel manager might find helpful. Information on publications from Prentice-Hall about labor relations is available directly from the publisher at Englewood Cliffs, NJ 07632.

PURCHASING

The purchasing agent orders all equipment, supplies, and materials used in the laboratory and provides liaison between the departments involved when large equipment is to be installed. The purchasing agent maintains ongoing communications with shipping and receiving, industrial engineering, and maintenance departments. When large items of equipment are ordered, everyone who will be involved in handling, installing, maintaining, and using them must be kept fully informed from the time the order is placed until the item arrives and is installed and working. The loading dock or delivery point must be specified to the vendor and the carrier. The building engineer must approve the site where equipment is to be installed. As a matter of fact, the building engineer should be consulted before any large, heavy item of equipment is ordered, about the selection of a site for its placement—a location free of any elements that may interfere with the operation of the equipment, and one that is not adjacent to laboratories that might be adversely affected by the operation of the new apparatus. There should be a standard guide designating which approvals are required for purchases within specified amounts. No item should be ordered until all departments or individuals concerned with its use have been consulted and approval is given by the designated authority.

The purchasing agent maintains very close coordination with the stockroom manager in order to maintain supplies at the appropriate level.

Personnel Requisition Form

Complete in detail and forward original to the Personnel Office

Title of vacant position	Position #	Budget line item #	Employing Department	Date

Full-time ( )   Part-time ( ) Indicate hours per week _____
Temporary ( ) Length of time (Months, dates, hours per week) _____
Indicate work hours/schedule if other than 8 to 5 Monday-Friday _____

Day shift ( )
Night shift ( )
Variable ( )

Base salary $ _____
Flexibility above base:
Yes ( ) No ( )

Educational requirements:

Experience requirements:

Other qualifications/skills required to perform this job such as: Typing wpm _____   Type of typewriter or word processor/personal computer _____
Adding machine/calculator: Yes ☐ No ☐   Dictaphone: Yes ☐ No ☐   CRT: Yes ( ) No ( ) entry ☐ retrieval ☐ both ☐   Others (Be specific in explanation, e.g., types of lab equipment) _____
Will training be provided on specific equipment Yes ( ) No ( )

Describe in detail job duties:

The areas listed below have been identified by Environmental Health and Safety as high risk areas. Will this person be working with or around any of the following: Indicate (X) if yes: Chemical carcinogens ( )   Recombinant DNA ( )
Oncogenic viruses ( ) Infectious Agents ( ) Animals ( ) Human blood/tissues ( ) Radioactive materials ( ) Experimental Drugs ( )   **If "Yes" is indicated on any of the above, please complete reverse side of this form.**

Immediate supervisor of position

Person(s) authorized to request job offer

Refer applicants to (Name, Title, & Phone #)	Alternate (Name, Title, & Phone #)	Days & hours interviews can be scheduled

Room # or address

Date position available

( ) New Position - Account # _____
( ) Replacement position - for whom _____

Account Title _____
Reason for leaving _____   Effective _____

ALL REQUISITIONS NOT COMPLETED IN DETAIL WILL BE RETURNED

All employment-related activities including referrals for employment are made without regard to race, color, religion, sex, national origin, age, handicap, or veteran status. Employing departments must make their applicant selection on an equally non-discriminatory basis.

Department Head's signature _____

APPLICANT'S NAME	REF. BY	DATE	RACE & SEX	RESULTS (REMARKS)

BELOW FOR PERSONNEL OFFICE USE ONLY

Date posted _____ Posting period up: _____ Requisition # _____

Hazard Codes:
C = Chemicals   V = Viruses   D = Recombinant DNA   R = Radioisotopes   A = Animals   H = Human tissue (blood, etc.)   E = Experimental drugs   I = Infectious agents

1. Which are used in the laboratory in which this person will be working? [Enter codes(s)] _____
2. With which will this employee be directly working? [Enter codes(s)] _____
If R is coded: 1. Name and Social Security number of investigator licensed to handle radioactive materials _____

    2. Social Security number of immediate supervisor _____
    3. Laboratory location _____

It is helpful if the stockroom manager and purchasing agent are physically close enough to communicate easily and frequently. Computers have greatly simplified the maintenance of stockroom inventories, and they can be programmed for automatic reordering when items reach a certain inventory level.

An imaginative purchasing agent will devise systems for expediting delivery of orders. Open purchase agreements with local vendors whereby orders can be placed by telephone and delivered immediately are a boon, almost a necessity, for obtaining items needed unexpectedly and quickly. Another useful system is to place orders by telephone, giving the vendor an assigned purchase order number. Formal paperwork can be put in the mail after the vendor has started to process the order. These are standard arrangements that are very useful; still, occasional errors are inevitable, so they must be carefully monitored.

The purchasing office must have a "tickler file" indicating when items are expected, in order to be alerted to delays. That file, constantly updated, can be placed where it is accessible to anyone who wishes to check on the status of an order. This system eliminates a lot of raised voices and demands for explanations.

The purchasing agent serves as a valuable consultant to scientists by keeping informed about the latest materials, equipment types and models, as well as supplies, as they come to the marketplace. He can advise on reputable manufacturers, the capability of certain instruments and models, and make suggestions of the item that might meet the scientific need.

Many scientific and technical magazines provide current information on the latest instrumentation. A resourceful purchasing agent should subscribe to (or be on the routing list of) those journals that offer such information. Each week, for example, *Science*, the publication of the American Association for the Advancement of Science, publishes a column entitled "Products and Materials," itemizing announcements of new apparatus. Similarly, *Nature* offers a column titled "New on the Market." Several journals publish annual guides to sources of scientific instrumentation, naming suppliers and manufacturers and giving other pertinent data. *Physics Today* publishes a world directory of physics research instruments, supplies, and services each August. *Physics Today* also publishes a comprehensive guide to science and technology instrument directories. An updated version of the August 1986 review of directories is shown on pages 39–42.

Purchasing agents should have the latest buyer's guide for their laboratory's specialty on their shelves for quick reference. They should also maintain a library of complete and current catalogs from major equipment manufacturers who supply items of interest to their laboratories so that researchers may consult them as needed. Catalogs offer the most recent models, specifications, prices, delivery dates, and other details. Purchasing agents should see that their names are on the mailing lists of key suppliers so that they are alerted to new products, price changes, or

PRODUCT GUIDE DIRECTORY*

AMERICAN LABORATORY BUYERS' GUIDE. Covers analytical instruments and includes a section on biotechnology. Published annually in January by *American Laboratory*. International Scientific Communications, 30 Control Dr., P.O. Box 870, Shelton, CT 06484. U.S. edition, $15 plus postage (prepaid); international edition, $17 plus postage (prepaid).

ANALYTICAL CHEMISTRY LABGUIDE. Covers chemicals, instruments, equipment, supplies, and research and analytical services. Published annually in August by *Analytical Chemistry*. American Chemical Society, 1155 16th Street, NW, Washington, DC 20036. $9.

ATE, INSTRUMENTS AND TEST SERVICES BUYERS' GUIDE AND DIRECTORY. Also covers software and work stations. Published each summer by *Electronics Test*. Miller Freeman Publishing, 1050 Commonwealth Avenue, Boston, MA 02215. $15.

AVIATION WEEK & SPACE TECHNOLOGY BUYERS' GUIDE. Covers products and services in the aerospace industry. Published annually in October by *Aviation Week & Space Technology*. McGraw-Hill, 1221 Avenue of the Americas, New York, NY 10020. $24.95.

CHEMCYCLOPEDIA. Covers chemicals, trade names, packaging, and applications. American Chemical Society, 1155 16th Street, Washington, DC 20036. $40.

CHEMICAL ENGINEERING EQUIPMENT BUYERS' GUIDE. Covers equipment, services, and engineering materials. Published annually in July by *Chemical Engineering*. McGraw-Hill, 1221 Avenue of the Americas, New York, NY 10020. $20.

CHEMICALWEEK BUYERS' GUIDE. Covers chemicals, raw materials, and packaging and transport of chemicals. Published annually in October by *Chemicalweek*. McGraw-Hill, 1221 Avenue of the Americas, New York, NY 10020. $20.

CONTROL PRODUCTS SPECIFIER. International coverage of control instrumentation, equipment and systems, automation, and robotics. Published annually in November by *Control Engineering*. Cahners Publishing, 1301 South Grove Avenue, P.O. Box 1030, Barrington, IL 60010. $15.

DATA ACQUISITION AND RECORDER HANDBOOK AND BUYERS GUIDE. Covers instruments and control systems. Published by *Measurements and Control*. Measurements and Data Corporation, 2994 West Liberty Avenue, Pittsburgh, PA 15216. $15.

(continued)

Product Guide Directory* (*Cont.*)

DATAPRO'S DIRECTORY OF MICROCOMPUTER HARDWARE (2 vols.). Covers microcomputers, software, and peripherals. Updated monthly. Datapro Research Corporation, 1805 Underwood Boulevard, Delran, NJ 08075. Subscription, $661/year.

DATA SOURCES (vol. 1: Software; vol. 2: Hardware/Data Communications). Covers microcomputers to mainframes, peripherals, terminals, data communications and telecommunications equipment, software, and services. Published semiannually in April and October. Ziff-Davis Publishing Co., One Park Avenue, New York, NY 10016. Subscription, $440/year; single set, $250.

EEM/ELECTRONIC ENGINEERS MASTER (4 vols.). Covers electronic components, equipment, and computers. Published annually in August. Hearst Business Communications, 645 Stewart Avenue, Garden City, NY 11530. $75.

ELECTRONIC DESIGN'S GOLD BOOK. Covers electronic components and equipment. Published annually in the fourth quarter by *Electronic Design.* Hayden Publishing, 10 Mulholland Drive, Hasbrouck Heights, NJ 07604. U.S. issue (6 vols.), $70; international issue (1 vol., 2 parts), $70.

ELECTRONIC INDUSTRY TELEPHONE DIRECTORY. Covers electronic components, services, and computers. Lists over 12,000 manufacturers and importers, 3,750 representatives, and 4,000 distributors. Published annually in August. Harris Publishing, 2057-2 Aurora Road, Twinsburg, OH 44087. $45.

ELECTRONICS BUYERS' GUIDE. Covers electronic components and equipment. Published annually in June by *Electronics.* McGraw-Hill, 1221 Avenue of the Americas, New York, NY 10020. $40 USA and Canada.

FIBEROPTIC PRODUCT NEWS BUYING GUIDE. Covers fiberoptics and related products, services, and manufacturing. Published annually in July by *Fiberoptic Product News.* High Tech Publications, 23868 Hawthorne Boulevard, Torrance, CA 90505. $55.

GUIDE TO SCIENTIFIC INSTRUMENTS. Covers research apparatus, materials, and services. The 1987 edition emphasizes biotechnology products and instruments. Published annually in February by *Science.* American Association for the Advancement of Science, 1333 H Street NW, Washington, DC 20005. $16 ($17.50 by mail).

ISA DIRECTORY OF INSTRUMENTATION. Covers instrumentation, automation, process management, and control. Pub-

(*continued*)

PRODUCT GUIDE DIRECTORY* (*Cont.*)

lished annually in January. Instrument Society of America Services Inc., P.O. Box 12277, Research Triangle Park, NC 27709. Price to be announced.

LASER FOCUS/ELECTRO-OPTICS TECHNOLOGY BUYERS' GUIDE. Covers laser, optical, fiberoptical, and optoelectronic equipment and devices. Published annually in January by *Laser Focus/Electro-Optics Technology.* PennWell Publishing, 119 Russell Street, Littleton, MA 01460. $65.

LASER AND OPTRONICS BUYING GUIDE. Covers lasers, optics, optronics, imaging, infrared, and some fiberoptics. Published annually in December by *Lasers and Applications.* High Tech Publications, 23868 Hawthorne Boulevard, Torrance, CA 90505-5908. $60.

MEASUREMENTS AND CONTROL HANDBOOK AND BUYERS GUIDE. Within each of the six issues published annually, a guide section covers transducers, instruments, and control systems for temperature, flow, pressure, and other variables. Published bimonthly in *Measurements and Control.* Measurements and Data Corporation, 2994 West Liberty Avenue, Pittsburgh, PA 15216. Subscription, $22/year.

MICROCOMPUTER MARKET PLACE. Covers microcomputers. Published annually in November. R. R. Bowker, 205 East 42nd Street, New York, NY 10017. $95.

MICROWAVES AND RF PRODUCT DATA DIRECTORY. Also covers fiberoptical and high-speed digital equipment. Published annually in October by *Microwaves and RF.* Hayden Publishing Company, 10 Mulholland Drive, Hasbrouck Heights, NJ 07604. $25.

NUCLEAR NEWS BUYERS GUIDE. Covers nuclear products, materials, and services. Published annually in March by *Nuclear News.* American Nuclear Society, 555 North Kensington Avenue, La Grange Park, IL 60525. $64.

OPD CHEMICAL BUYERS DIRECTORY. For the chemical, plastics, coatings, oils, and drug industries; covers chemicals, shipping, transportation, and terminals. Published annually in October by *Chemical Marketing Reporter.* Schnell Publishing, 100 Church Street, New York, NY 10007. Available only by subscription to *Chemical Marketing Reporter,* $65/year.

THE OPTICAL INDUSTRY AND SYSTEMS PURCHASING DIRECTORY (vol. 1: Photonics Buyers Guide; vol. 2: Photonics Handbook; vol. 3: Photonics Dictionary). Covers lasers, optics, fiberoptics, electro-optics, and imaging. Published annually in

(continued)

PRODUCT GUIDE DIRECTORY* (*Cont.*)

March. Laurin Publishing, Berkshire Common, P.O. Box 1146, Pittsfield, MA 01202. $76, three-volume set (prepaid).

RESEARCH & DEVELOPMENT 1987 TELEPHONE DIRECTORY. Covers laboratory apparatus, instruments, chemicals, components, and materials. Published annually in February by *Research & Development.* Cahners Publishing, 1301 South Grove Avenue, P.O. Box 1030, Barrington, IL 60010. $15.

RESEARCH SERVICES DIRECTORY (3rd ed., 1987). Covers 3,500 firms, laboratories, individuals, and other facilities in the private sector that provide contract or fee-based research services. Supplements published periodically. Gale Research Company, Book Tower, Detroit, MI 48226. 1987 edition, $285; supplements, $150.

SEMICONDUCTOR INTERNATIONAL MASTER BUYING GUIDE. Covers chemicals and materials, production and processing capabilities, measurement and testing, and services. Published annually in the first quarter by *Semiconductor International.* Cahners Publishing, 1350 East Touhy Avenue, P.O. Box 5080, Des Plaines, IL 60018. $25.

SOFTWARE CATALOG: SCIENCE AND ENGINEERING (4th ed.). Derived from the *International Software Database.* Covers more than 4,300 programs for micros and minis. Published annually in April. Elsevier Science Publishing, 52 Vanderbilt Avenue, New York, NY 10017. $65.

SOLID STATE TECHNOLOGY PROCESSING & PRODUCTION BUYER'S GUIDE. Covers materials, equipment, and services used in the manufacture and processing of solid-state devices. Published annually in February by *Solid State Technology.* PennWell Publishing Company, 14 Van Deventer Avenue, Port Washington, NY 11050. $50.

TEST AND MEASUREMENT WORLD ANNUAL BUYER'S GUIDE. Covers test, measurement, and inspection equipment; software; computer-aided-design equipment; microelectronics; and rf and optoelectronic equipment. Published annually in July by *Test and Measurement World.* Cahners Publishing, 275 Washington Street, Newton, MA 02158. $4.

*Updated from *Physics Today*, August, 1986.

announcements of apparatus failures that may require engineering changes in instruments currently in use at their institutions.

Although it is the scientist who must make the final decision about ordering instruments and equipment for use in research, an intelligent, well-informed purchasing agent who is enthusiastic about the work soon earns the respect and confidence of the research staff and contributes much, albeit indirectly, to the scientific output of the institution.

COMMUNICATIONS SERVICES

The communications services of the laboratory—internal and external publications, conferences, public relations, staff meetings—are under the direction of the research services manager, who selects the personnel and oversees the various functions. The head of the laboratory, however, is intimately concerned with all internal and external communications and will therefore work closely with the editor, the public-relations specialist, and the staff person responsible for the arrangements of meetings and conferences. It is strongly advised that the final decisions on key appointments in the communications area have the approval of the head of the institution.

In addition to the usual professional qualifications requisite for performing the functions that comprise "communications," candidates for positions should be evaluated on their awareness of the special problems that may arise in the dissemination of scientific information. They must be willing to defer to the researcher in every situation having to do with the presentation of that individual's work to the public or to the scientific community and must understand why it is that sometimes even exciting news cannot be immediately broadcast. A good rule of thumb to follow is that internal information flows on a green light, and external information proceeds on a yellow light of caution. It is essential that everyone concerned with the release or public dissemination of information from the laboratory has clear guidelines concerning the approvals that must be given prior to any such release. The head of the laboratory determines the hierarchy of approvals required for particular kinds of subject matter. The public-relations specialist will, in all likelihood, establish contacts with reporters and writers in the news media who will call the public-relations office when they are seeking information, and it is the public-relations office that prepares and disseminates routine news releases when there is something to be reported.

The laboratory director is ultimately responsible for any mention in the media of the institution and must take great pains to ensure that those responsible for the release of information are sensitive to the necessity for giving out accurate information, and for always obtaining proper authorization for its release. If something appears that creates a false impression or arouses protest and resentment from the public, it will be the laboratory director who bears the brunt—not the public-information or public-relations office. The quality of the communications staff will have a telling effect on the way the laboratory is viewed in the community and no effort should be spared to select those who are well qualified for the job professionally and also exceedingly astute in handling assignments.

All administrative activities are susceptible to Parkinson's Law, but perhaps none so much as those that fall into the general category of communications. Community relations, publications, and organization of special events are open-ended functions that can go as far as the resources permit, and the work does, indeed, expand according to the people and resources available. Therefore, this function must be carefully monitored to ensure that the number and kinds of specialized personnel, such as editors, photographers, artists, and writers, does not grow beyond the limit that is desirable. The main public-relations representatives of the laboratory are the scientists, and chief among them is the head of the institution. The role of the communications staff is to assist the scientific staff in performing that role in the manner and to the degree determined by the policies of the organization.

FACILITIES AND MAINTENANCE

The physical atmosphere of the laboratory requires knowledgeable oversight by a trained professional who can deal with planning new facilities and cost-effective renovations, waste-disposal systems, air-supply testing, noise and vibration problems, and identifying sources of air pollution and making recommendations for their segregation or elimination. The plant engineer may also supervise the general maintenance staff—janitors, groundskeepers, pipefitters, electricians, carpenters, painters, and so on—and oversee the work of the crafts that service the scientific staff— glassblowers, machinists, welders, electronics specialists, and so forth.

The plant engineer who provides and maintains a working environment that is physically comfortable and who meets the criteria for carrying on research also contributes enormously to the morale of the institution. The confidence that day-to-day building problems and minor maintenance needs will be taken care of promptly and competently is an impossible-to-overvalue boon to those engaged in the frustrating search for new knowledge in the world of the physical, biological, engineering, and social sciences. The person selected for the position needs a certain amount of technical training and, in addition, must have the interest and enthusiasm to keep up to date on new ideas, materials, equipment, and systems that introduce the possibility of new solutions to old problems and shed light on better ways to handle various aspects of plant operations.

As with other nonscientific positions in a laboratory, this one has some serious drawbacks, a major one being that it is the top job in its category within the institution. In a sizable laboratory, someone may rise to the position after years of assisting in a particular aspect of engineering maintenance or craftswork, bringing to the post a thorough knowledge of the physical facilities as well as acquaintance with the staff. Those who have acquired both maturity and seniority by the time they reach the top position are more likely to be satisfied in the job than are those brought in from the outside. Given the complicated supply and service systems and the multifarious needs of a research laboratory, it takes time to become familiar with the idiosyncrasies of any structure. By the time plant engineers become well acquainted with their responsibilities, they may already be restless to move on, unless there are stimulating challenges to test their ingenuity. Fortunately it is not usually a problem to provide these challenges. A research laboratory is not a static piece of architecture, but must be capable of progressive remodeling inside; with the passage of time, old needs become outdated and new needs appear in even the most beautifully designed laboratories. The physical growth and alterations in space arrangements or addition of new systems and discard of old ones occur as research programs grow, change, or are superseded by others. The plant engineer who can be excited and stimulated by the challenge of solving problems, can come up with new solutions or new materials or a new way of combining both to serve a purpose, and is willing to make an effort to keep abreast of developments in the laboratory facilities field is ideal for the position.

A bimonthly publication for facilities managers available by subscription is *Facilities Planning News*. (For subscription information, write to: P.O. Box 1568, Orinda, CA 94563.) *Facilities Planning News* also mails out, upon request, information on special conferences. In 1986, special-conference mailings were available from meetings dealing with R&D facilities, facilities renovations, high-tech facilities, office projects, new plans for facility flexibility, and other specialized topics.

REFERENCES

1. Raju, M.R., Phillips, J.A., and Harlow, F. (eds.) *Creativity in Science—A Symposium* (LA–10490–C). Los Alamos, NM: Los Alamos National Laboratory, 1985. p. 98.
2. Balderston, J. Successful Administration of a Research Laboratory. *Research/Development* 20(6):24–30, 1969.
3. Balderston, J., Birnbaum, P., Goodman, R., and Stahl, M. *Modern Management Techniques in Engineering and R&D.* New York: Van Nostrand Reinhold Co., 1984.
4. Weisskopf, V.F. The Frontiers and Limits of Science. *American Scientist* 5(4):409, 1977.

Chapter 4

RESEARCH STAFF

. . . [science] is predestined by the laws of logic and the nature of human reasoning . . . almost as though it had once existed, and its building blocks had then been scattered, hidden, and buried, each with its unique form retained so that it would fit only in its own peculiar position.

—Vannevar Bush, *Endless Horizons*

T he creation of a distinguished scientific staff may be compared with the way in which random rubble walls (seen in parts of England and almost everywhere in Scotland) were built: irregular pieces of stone loosely packed, the interstices between the large stones filled by small ones, and set dry without mortar. One marvels at their durability and ponders on the judgment that went into selecting each stone to rest securely on previous ones, to strengthen those surrounding it, to provide a foundation for those that follow. So it is with great research teams. Their strength arises out of the particular qualities of each individual positioned in the right place at the right time in relation to others so that their combined force becomes an entity whose power is greater than the sum of its parts.

The laboratory's research program is inseparable from the people who comprise its research staff. Just as writers are the currency of publishing houses and actors the cash flow of theatrical agencies, so scientists are the capital of research laboratories.

APPOINTMENTS

The techniques used for bringing new people into your research organization are various and differ widely depending upon the reasons for the recruitment and the contributions newcomers are expected to make. From time to time, appointments are made to every laboratory's research staff to augment the work of an ongoing program, to stimulate the research environment by bringing in young scientists at the postdoctoral level, or to replace key persons in major and continuing projects. Approaches to finding applicants in these categories differ from those that are effective in enticing senior scholars to sever their connections with one institution and take up their work at another laboratory.

Recruiting Senior Scientists

Persuading a recognized leader in a particular field to leave one place and move to another is, in the simplest analysis, a selling job, and is always accompanied by a great deal of negotiation. Questions of space, equipment, budget allocation, and various perquisites must all be discussed with openness and candor. The candidate who is being "wooed" sometimes makes demands that are clearly unreasonable—the temptation is great—and through overeagerness to win the suit, the "wooer" might agree (or seem to agree) to things that may never materialize. Successful recruiting campaigns, involving lengthy negotiation, may often be followed by some disappointments on both sides, and every effort should be made to restrict those to relatively minor matters. Major questions having to do with funds, personnel, space, instruments—these are critically important. If the parking space assignments have to await the new lot, or the local schools and recreational facilities are less irreproachable than was reported, new arrivals will probably be able to take those disappointments in stride.

Negotiations are likely to be even more complicated by the fact that major scientific researchers rarely travel alone—they nearly always move with assistants and coworkers who have been associated with them for some time, and these people must be included in the recruitment plan. Therefore, the selling job in some cases extends to the staff of the scientist who is being recruited. It is very important, however, that the laboratory's standard personnel procedures and policies are strictly applied to all employees brought in by this arrangement, particularly in the establishment of salary levels. They should be commensurate with what is being paid to others with similar education, experience, and responsibilities.

Successful recruiting does not allow you to mislead candidates in any substantive matter, but it does allow you to emphasize the positive aspects of the institution, its geography, its research facilities, and the scientific resources and climate in the community or nearby.

Commenting on recruiting scientists in government laboratories, Frederick J. de Serres at the National Institute of Environmental Health Sciences (NIEHS), Research Triangle Park, North Carolina, noted:

> Successful recruiting in a government laboratory can be difficult, but in Research Triangle, we have no problem. Duke University, UNC-Chapel Hill, Wellcome Research Laboratory, and other institutions make this a wonderful place to carry on scientific research. The NIEHS labs are in a beautiful wooded area, the climate is wonderful.
>
> Of course we have to work at it. It depends a lot upon the way an interview is handled, not only from the professional point of view but from the social and personal as well. In this scientists are just like anyone else who might be considering a family move; they want to know about living conditions, schools, availability and cost of housing, and so

forth. They also want to know what kind of atmosphere exists in the laboratory and it is very important to have them meet their prospective colleagues in a social group as well as in the laboratory. We always meet candidates at the plane, transport them to the lab, and arrange for them to see as many of the scientific and other staff members as possible.

Enticing recognized scientists to join your staff is difficult, but at least it has the advantage that the object of the campaign is a known element, and his or her worth to your laboratory can be assessed. Today's "stars," today's prize winners, are known; but creative research directors aspire to the vision that enables them to perceive the prize winners of tomorrow. Unfortunately, young researchers do not come branded with visible marks of future eminence or signals that identify them as those whom Vannevar Bush described in *Endless Horizons* as "men of rare vision who can grasp well in advance just the block that is needed for rapid advance on a section of the edifice to be possible, who can tell by some subtle sense where it will be found, and who have an uncanny skill in cleaning away dross and bringing it surely into the light."[1]

Recruiting Postdoctoral Fellows

One of the best ways of discovering budding researchers who will earn future prominence is through an active postdoctoral program. Such programs are designed to give full-time research experience to young scientists whose only independent work up to that time has been Ph.D. thesis research, which isn't always as independent as one supposes. Moreover, graduate students often have teaching responsibilities up to the completion of their doctoral programs and have never worked at research to the exclusion of everything else. Usually appointments are made for a finite period—an initial one-year term with a possibility of renewal for a second or even third year. These plans give fellows a year to demonstrate their industry, enthusiasm for research, and way of thinking. Often, it allows enough time to plan a research project, assemble the necessary equipment, and get started. But it is rarely enough time to carry out a substantive piece of work. When indications are favorable, the appointment can easily continue; but if things don't quite work out, the appointment can be terminated at the end of the year without embarrassment on either side.

Postdoctoral programs may be funded from internal or external sources, but in order to present a good case to obtain funding, plans must be carefully mapped out. An important element is the method of selecting fellows, and the senior research staff should participate actively in this process. A selection committee, composed of representatives of major research programs, might recruit and screen candidates and present their final recommendations to the laboratory director. It is essential that all significant research areas be represented on the committee. Two reasons are paramount: (1) the most effective way to recruit superior candidates is through the university departments where members of the research staff received their training or have professional acquaintances; and (2) it will help to ensure that every research area in the laboratory has an equal opportunity to bring in fellows.

The cooperation of the senior research staff is important to the success of the program. It will not only assist in recruiting fellows of high caliber, but it is they who will serve as advisors, mentors, or group leaders for the fellows, and, at least for the first year, it is they who are to take time out to talk with them, observe their approach to research, and assist in evaluating their work at the end of the fellowship period.

The principal objective of such a program, in its broadest sense, is to nurture scientific talent and pass on to the next generation the wisdom and experience that exist within the senior staff of your institution. Every scientific laboratory has the responsibility to contribute this much to science. A by-product may be the development of a valuable member of your future staff. George Beadle became the head of the laboratory where he first worked as a postdoctoral fellow, the biology department of the California Institute of Technology. And Howard I. Adler, who went from Cornell to the biology division of the Oak Ridge National Laboratory as a postdoctoral fellow, later became the division director in the years following Alexander Hollaender's retirement. But, as a rule, it is advisable for young scientists to gain experience in several laboratories, staying in each place long enough to accomplish something substantial—usually a piece of work that results in publication.

The standard application form for research personnel will not serve for postdoctoral fellows; a special one must be designed to include the items of particular interest to your postdoctoral committee. Questions that are irrelevant should be omitted. The committee must have a full resume on each candidate, which includes vital statistics; details about education, research interest activities, and future plans; and references who are knowledgeable about the candidates' scientific interests, accomplishments, and potential. Some laboratories ask postdoctoral candidates to include letters of reference in the application package; others prefer to request them on a confidential basis. In whatever manner the letters are obtained, it takes a special technique to read them. Experience helps, and some acquaintance with the letter writer is an invaluable guide. Knowing that a particular professor is a hard grader, for example, one may presume that praise given even to top-flight candidates may be restrained. And the reverse is also true; if the writer is known to speak well of everyone and only in glowing terms of former students, a lukewarm letter might be read as highly derogatory. The best clues in reference letters are found in the omissions, that is, what did the referee *fail* to say? Letters may state that candidates are intelligent, hardworking, conscientious, sincere, and other wonderful things, but never mention whether they ever had a good idea. At this point, face-to-face or telephone inquiries usually produce fuller, and frequently more honest, evaluations than those contained in letters.

Some people are reluctant to mention negative attributes of others in writing.

The laboratory head does well to consult the senior staff in cases where doubts arise. An inquiry may reveal that a highly qualified candidate is considered to be difficult, hard to get along with, not generally liked by colleagues. If a team worker is being sought, the decision is easy. But if originality and brilliance are wanted, and the candidate shows evidence of being "just the block that is needed," the director might choose to take a chance, even against the recommendation of the senior staff. Many have done so to their regret; others have lived to boast of their decision for the remainder of their careers.

Recruiting of postdoctoral fellows and announcement of staff openings must conform to federal and state regulations designed to provide equal opportunity to candidates, and notices placed in carefully selected professional publications must clearly state that the regulations are being complied with. The postdoctoral committee, with its representation from various fields, can advise on the most suitable publications. For advertising position openings, those journals with the widest circulation, such as *Science* and *Nature*, are good choices. For recruitment within the academic community, announcements should also be placed in the *Chronicle of Higher Education*. Examples of advertisements are shown on page 51.

Printed flyers, or even typed notices, should be placed on bulletin boards at scientific conferences and meetings attended by professionals in the appropriate fields, calling attention to openings or postdoctoral appointments. Announcements of postdoctoral programs are always sent to graduate schools of the sciences, but their effectiveness is somewhat eroded by the surfeit. Bulletin boards become so overloaded that older flyers get buried as new flyers arrive.

Some annual meetings of scientific and professional societies provide opportunities to interview candidates for openings. Frequently, interview rooms are set up where members of your staff can meet face-to-face with scientists seeking employment. Each society handles this differently, so it is a good idea to inquire long before each annual meeting about announcing your openings, what (if any) fees are required, and other details.

One of the best methods of identifying superior candidates for postdoctoral fellowships is through the system known as "the old-boy network"—an approach that fell out of favor when affirmative action and other equal-rights laws were enacted. Asking your former professor or a close professional associate to "encourage your best students to send us an application," is about as far as one can go now without jeopardizing your institution's position. One research director at a noted institution cast a new light on the old-boy system:

> We have been hampered by the assumption that we operate only through the old-boy network, and good people did not apply, assuming that if they had not been suggested by someone in the network, it was

RECRUITMENT ADVERTISEMENTS

hopeless. What this created was a situation in which only third-rate people applied from the outside, occasionally a second-rater. The first-raters were snapped up by other places, and we were *forced* to resort to the old-boy network to get good people. This is now changing; of our last three fellow appointments, two were outsiders who had applied in response to our ad, and one was an insider. It is not our efforts that have brought about the change; the job market is so bad that we are even getting applications from *very good* people who may not be in the network!

A deadline must be set for receipt of applications, and notices must include a name and address from which application forms and full details are available.

Recruiting Midlevel Scientists

Scientists who are past the postdoctoral level but who have not yet reached the senior-scientist category may be recruited by using some of the same procedures as for selecting postdoctoral fellows, but there are additional techniques that may be employed.

Roy Curtiss III of Washington University, St. Louis, who has had extensive administrative experience, makes the following suggestion:

> The recruitment of an outstanding, established, young investigator still on the way up is readily straightforward if the research laboratory has on its staff one or more individuals of international stature in a field closely allied with that of the individual being recruited. In other words, the existence of people of eminence in a given area will make it easy to attract others of increasing eminence in the same area.
>
> On the other hand, if it is decided by the laboratory management to branch out into a new area and to recruit a more senior, well-established individual to help build that area, then that is another matter—especially if there are no members of the staff with a commanding knowledge of this new area to be developed. In this case, it might be appropriate to invite four, five, or six of the most distinguished scholars in that area to visit the institution and present a seminar and to review plans for building up research in their area of expertise. In most instances, the individuals invited would themselves be unavailable, and may not even be desirable candidates because of their seniority. They can, however, help to identify promising younger scientists and even serve as consultants for evaluating credentials of applicants when the search begins.
>
> When a candidate has been selected, these individuals may also assist in the recruitment effort by counseling and perhaps persuading the candidate to accept the offer.

It is imperative that candidates for midlevel positions visit the laboratory and lead a seminar where they meet their potential colleagues and where their compatibility with the work of the laboratory and the other members of the staff can be assessed.

Visiting Scientists

Temporary or short-term researchers are like houseguests; they are quite a bit of trouble, but are usually worth it. Outsiders present a challenge to which hosts must rise, so that their home institution appears in the best possible light. You want to show off its most admirable aspects, and to make the visitors' sojourns both pleasant and stimulating. For their part, the visitors bring fresh points of view, new information, and usually a flattering curiosity about the activities of their hosts. If nothing else, they are almost always good listeners.

R. Christian Anderson, when he was assistant director of Brookhaven National Laboratory, said, "A research center keeps fresh only to the extent that it has a high turnover of staff or has a mechanism whereby a large number of visiting scientists find it attractive to spend part of their time in residence."

High turnover is expensive and not conducive to long-range development, but programs designed to attract a steady flow of guest researchers are simple to administer and relatively inexpensive. Laboratories with special expertise or equipment attract visitors who are willing to come on their own funds. And programs for training may be set up with grants from public or private sources. College and university teachers whose research time is limited during the school year often seek out places where they can spend a summer performing full-time research or catching up on newer work in their special fields. Many are also able to find grant support for such brief visits. Laboratories that fund their own visiting programs can advertise them and attract a broader range of candidates. Every scientific institution has the responsibility to provide training in some form, and every laboratory can benefit from exchange of information with researchers from other places. The benefits are more likely to be mutual if the visitors are carefully selected so their interests, talents, and abilities are compatible with the scientific goals and level of work being done in the host institution.

How to Read a Resume

Just as the means of recruiting personnel at different levels are various, the techniques of evaluating the material included in the resume or the application form also differ. The resume of an eminent scientist who is being lured to the laboratory and whose work is well known will be used mainly to introduce him or her to seminar groups and to make an announcement of the appointment to the press when and if it occurs.

Postdoctoral candidates will be assessed on the information in the application package that may indicate potential for development into a first-class scientist. Applications for staff positions that result from announcements of openings must be assessed in relation to the specific post to be filled.

It has become a common practice in preparing resumes to enhance one's background and even to exaggerate in some cases. In fact, that concept has spawned a "professional" group—people who, for a fee, prepare resumes and promise to make the candidate more attractive by the manner in which the information is presented. The professional resume is usually easy to spot because it always looks better than one prepared by the candidate. It is often printed with two or more type faces; it is well placed on the page, balanced so all the information is not on one side; and generally there are no misspelled words. Such resumes should be approached with caution. The resume specialist can become overenthusiastic, and it is a fine line that separates the bare facts from the decorated version of some entries. A resume may be completely factual, may include all pertinent data, but—and this applies to all resumes no matter who prepares them—one must be able to read between the lines, and try to guess what has been omitted. In many cases, those literary curlicues do not affect overall content of the resume, but it is well to be aware of their presence or possible presence. Decoration is one thing; disguise is another.

One is much less likely to encounter inflated resumes in science than in the business world, because the publications list is a major element in the resume of a scientist, and that cannot be disguised short of outright falsification. This occurs so seldom that it scarcely bears mentioning. One has to be a fool to try it; there is always the chance that someone on the staff is sufficiently acquainted with the literature in the field to notice a fraudulent citation.

The list of references given on resumes can be assumed to be people who will speak well of the candidate. It is a mistake, however, to neglect to check out references on the presumption that they will not yield useful insights, keeping in mind that referees are more likely to give candid assessments of an applicant in conversation than in writing. Also, in conversation it is possible to press for further information on particular points that may be important in relation to the appointment being considered.

The Interview

Review of applications or resumes is the screening device used to select those candidates to be invited to visit for a personal interview. Invitations may be extended over the telephone to agree upon a time for the interview, but a formal written invitation must follow, giving the particulars: time, place, length of visit; arrangements for travel costs and living expenses; and documents or materials needed in addition to those already on hand, such as article reprints or copies of books. Except in unusual cases, the laboratory pays for the cost of travel in connection with interviews.

It is a good idea to include plans for some socializing. This need not be elaborate—it can be a backyard cookout, a hike in the mountains, or a small dinner party. The idea is to give applicants a chance to interact with their potential colleagues outside the laboratory setting.

As soon as a decision is reached to offer an appointment, the candidate should be notified immediately. If there is doubt as to whether the offer will be accepted, and if it is anticipated that further negotiation may ensue, the offer may be conveyed by telephone and then followed up in writing. Your written offer must be specific on such points as salary, starting date, fringe benefits (including health and life insurance), pension plans and retirement systems, vacations and sick leave, and other perquisites, as well as any points that were agreed upon during the negotiation.

Among the questions that frequently arise during employment consultations are those involving other commitments. Researchers may have part-time teaching posts or consultantships that they wish to retain. They may also hold positions in professional societies involving the use of laboratory time, facilities, and personnel, such as secretarial assistance. All researchers want some kind of understanding about travel for professional purposes, to present papers or merely to attend scientific conferences and professional meetings. The laboratory policy on travel reimbursement must be clear and the mechanisms for approval in advance of the travel set forth in unambiguous terms.

Obviously, every question cannot be anticipated and it always takes a while for a new fellow or staff member to settle in and become oriented. A brochure stating the general personnel policies can cover many of the points—the institutional policies on fringe benefits, vacations, sick leave, salary review, official travel, outside consulting, and similar topics. It may be possible for the personnel office to develop one booklet that will serve all personnel categories, or it may be necessary to have one for research staff and another for nonresearch personnel. This depends upon the variation in the policies, the size of the institution, and the details to be included in the publication.

STAFF RELATIONS

The group leader or other appropriate person should spend time with new staff members to acquaint them with the logistics of the laboratory; to describe the administrative policies; to explain about the chain of command through which certain requests and approvals are routed; to discuss the policies on performance reviews, promotions, and salary scales; and to answer any questions the new staff member may wish to ask.

Some large organizations provide periodic orientation courses to discuss institutional policies, but because of their scheduling, some staff members may not be invited to attend these sessions until quite some time after joining the staff. And often the material included in these briefings is not detailed enough to cover the unique aspects of each department or laboratory.

Departmental booklets containing general facts are helpful if filled with information about ordering equipment and supplies from outside vendors; obtaining stockroom items, secretarial, and editorial services; using official vehicles, restricted areas such as quarantine rooms or radiation sources, parking-lots, and computers; making long-distance telephone calls; and other relevant matters. Booklets should also give telephone numbers and the locations of technical and administrative service departments including the maintenance engineer; electrical, machine, and welding shops; purchasing office; library; stockroom; personnel office; cafeteria; editorial office; and medical dispensary. Emergency procedures should be clearly spelled out and the key telephone numbers listed even though some of them may also be affixed to every telephone and posted in every laboratory.

Performance Evaluation

When asked directly about the frequency of performance reviews, most laboratory directors reply, usually in a perfunctory way, "Once a year." Actually, review and assessment of the value of each individual's contribution to the goals of the institution should be going on all the time, and usually they go on whether the administrator recognizes it or not. But there must be a systematic procedure for performance reviews, and they should take place at least every twelve months to assess the progress of each staff member and to determine salary increases. Reviews may involve department heads, institutional executives, and others, and they must be done on schedule and care taken that no one is overlooked.

There are times, outside of scheduled reviews, when special situations arise—a job offer from another laboratory at a larger salary, for example—when compensation of a researcher must be assessed in relation to the contribution of that person to the laboratory's programs. Each case must be considered on its own merits, and, no matter what the circumstances, wise directors will not allow the spirit of competition to draw them into a bidding contest. Sometimes, the best reply to an announcement of a "better offer" is "Congratulations!"

Then there are those rare (and fortunate) times when a scientist or a research team comes up with an outstanding piece of work that propels them into sudden prominence. Offers from other institutions are sure to follow. The head of the laboratory should lose no time in discussing with those involved their future plans under the changed circumstances, including discussion of such things as salary, space, equipment, and other perquisites. Changes may sometimes include additional administrative responsibilities when a new group is set up or the staff is enlarged. But not all research scientists need or care for fancy titles or prestigious posts.

James B. Fisk, once president of AT&T Bell Laboratories, has noted,

"A contributing scientist should be made to feel no need for organizational tail feathers to attain personal satisfaction and financial reward commensurate with his contribution." Fisk was referring mainly to industrial science, but his remarks are appropriate in any research setting. "Every effort," he continued, "must be made to ensure that salary bears a direct relationship to performance and that the compensation of professional scientists is competitive with other related opportunities. Yet salary alone, however important and honorable and desirable, cannot be made to substitute for long for other factors which, in sum, constitute a good research environment."

Fisk's point is well taken. Outstanding scientists are more likely to be "bought" with superb facilities and research freedom than with an impressive salary.

Evaluating those outstanding scientists who are recognized as leaders in their fields is easy. And it is not difficult to identify those whose contribution does not meet the standards of the institution. There are many scientists whose work is solid, reliable, and valuable—but not in the same category as the "stars." Some may be marginally productive in research but make important contributions in other areas such as training and advising young researchers, performing key committee functions, serving as the major liaison with the surrounding community, organizing seminars, or holding national office in professional societies. These are all activities that contribute to the prestige and productivity of the laboratory.

The most obvious evidence of productivity is frequency of publication, but complete performance evaluation has to go further than counting papers. Some kinds of research lend themselves to short experiments that produce a spate of results in a brief period and a concomitant flow of articles, whereas, in other areas—some of the most significant—good papers come only after an extended period of research. William A. Arnold, a member of the National Academy of Sciences, famous for his research into the wonders of photosynthesis, used to say he felt good if he could bring out a paper every five years. It should be noted that he never published an insignificant paper. When Arnold received the American Society of Plant Physiologists' Kettering Award, the citation read, in part, "for his application of the rigorous principles of physics to a biological phenomenon; his intrepid manipulation of biological materials; and his ability to grasp a problem, frame it in experimental terms, and to follow its resolution full course."

The quality of the journal in which a paper appears deserves some consideration. Refereed journals are more selective than nonrefereed ones. Citation analysis—frequency of citation of an article and by whom and where cited—is believed by some to be a useful evaluation tool. In support of this, it has been pointed out that Nobel laureates usually have many more citations to their papers than do average scientists and that they also (but not in every case) had high citation ratings before they received their prizes.[2] The hazard of using citation analysis for performance evaluation is the same as for using it to select library journal subscriptions—articles and authors are cited for a variety of reasons, not all of them laudatory.

Objective measures may contribute something to performance evaluation, but the laboratory head has a great deal of subjective information to go on. One significant factor is the tendency of other researchers to consult a colleague. It was generally thought that William Arnold could well have produced more articles if he were not in such demand by other researchers for advice, guidance, and assistance with their problems, for which he was always graciously available. His broad knowledge—he was a physicist, a physiologist, and a biologist—made him a walking reference source for others, particularly young people with whom he was especially generous.

Some scientists are more theoretical than practical, and their contribution might be challenging others, stimulating them to consider original approaches to problems that do not yield to standard methods. At seminars, they may pose incisive questions that open up new avenues of thought to those whose work is being presented. Their own record of achievement in the form of publications may be low, but their influence in the laboratory may be immeasurable.

When Separation Is Inevitable

All research scientists go through fallow periods that usually prove to be temporary. The reason for this is not fully understood. Dry spells may follow highly productive periods, and sometimes they seem to result from frustration and discouragement or other reasons. If such a phase goes on for a long time, some kind of change may be called for. J. W. Boag, Royal Marsden Hospital, Sutton, England, says, "Everyone needs re-treading at 40. Why not a two-year sabbatical at that time to 'ginger' them up?" But even a sabbatical may not always work, and sadly but inevitably there are times when separation is unavoidable. There are also times when program focus or economic pressure may decree staff reductions, and as one popular song phrases it, "Breaking up is hard to do."

Personnel policies should take into account the fact that not all appointments are going to be entirely successful. As with postdoctoral fellowships, it is a good idea to make initial appointments for a finite period, with the understanding that they may be extended. Academia has done this for years with the tenure policy, but even that procedure does not solve all problems. Research moves on, people change, and sometimes the objectives of the institution change.

If budgetary stringencies can be cited, and particularly if more than one person is involved, it is somewhat easier, because the delicate issue of performance can be sidestepped. It is there, of course, since no chief executive elects to let the best people go even in the direst times. However, the decision may not revolve entirely around individual scientists but may be a matter of eliminating certain programs or areas of research. There is ample room in these situations, also, for a great deal of face-saving all around.

No matter what the reasons are (barring those rare cases of misconduct or criminal acts), it is never easy to let people go, but there are ways of doing it that cushion the blow. First of all, it should never be abrupt. It usually takes at least a year to find a research or teaching position, and often it takes longer. Every professional scientist deserves the consideration of that much time and more if possible. It should also be recognized that some scientists who flounder in one environment will thrive in another. There are myriad environmental factors that may become handicaps for any creative individual, and scientists are no exception. Some need the stimulation of students and do best in an academic setting where research and teaching are combined; some find teaching distracting and are most productive when they devote their efforts fully to research. One does better in isolation; another requires constant communication with colleagues in the same or peripheral fields. An analysis of the strengths and personality of an individual researcher suggests what environment may be conducive to increased productivity. The separation need not be acrimonious if the laboratory offers to help with the relocation by writing letters of recommendation and by nominating the departing staff members for positions in other institutions that may offer a more congenial setting for their talents.

It is exceedingly important for the morale of everyone on the staff that separations be handled with as much tact and consideration as possible, because the unspoken, perhaps unconscious, thought in everyone else's mind at such times is, "There, but for the grace of. . . ."

RESEARCH ENVIRONMENT

The formula for furthering creativity in a research-and-development organization given by S. J. (Sol) Buchsbaum, executive vice-president of AT&T Bell Laboratories is, "Hire the best, give them the best, demand the best." Of his own laboratories, he said, "We recruit and hire the best people available, fund them adequately, and motivate them to do their best work by providing a stimulating, challenging environment, continuing training, and educational growth opportunities. And, perhaps most important, we always demand and reward a commitment to and attainment of excellence."[3]

These standards are ambitious, but they represent an ideal to which others may aspire. "Hire the best" was even modified by Buchsbaum in the same statement when he revised it to "the best people available." But "demand the best" need never be qualified. This was the key to Alexander Hollaender's success as a research administrator. He believed his people were the best and imbued them with some of that feeling by demanding the best from them.

Research Assistants

Giving researchers the best does not necessarily mean elegant facilities and the latest instruments. The attractions of modern structures and state-of-the-art equipment can be seriously diminished by inadequate building and utilities maintenance, poor waste management, and incompetent technical and administrative service. Therefore, the admonition to hire the best is not limited to the scientific staff but should apply to everyone. And when achievements are being recognized it is important that the contributions of the supporting and service personnel are recognized. Recognition of good work at every level is a powerful morale builder and goes a long way toward unifying the efforts of the entire staff.

One branch of the research team that can easily be undervalued and overlooked when rewards are being handed out is the research assistants. Carl J. Sindermann says, "Scientists have known for a long time, but have been very careful to conceal the fact, that behind many successful careers are very competent and enduring laboratory technicians or research assistants (male or female)." Sindermann then asks, "Why then are scientists so often stupid or uncaring about such a critical part of the foundation of their success?"[4] He cites the qualifications of the ideal assistant as stated by a former mentor, Addison J. Pead, as "someone with a science background, a near-genius I.Q., training in equipment use and maintenance, computer expertise, typing skills at 120 words per minute, no outside commitments, no psychological hangups, and a perpetually positive outlook on life."

If such paragons were to be found within the fallible human race, they would indeed be ideal research assistants or any other kind of helpmeet, but most scientists feel fortunate to find an assistant with even a few of these qualifications.

Before beginning the search for a research technician, the principal scientist must have a clear idea of the qualifications that are essential to the position, those that are desirable but not absolutely necessary, and those qualities that must *not* be present. The personnel office may set general standards for employees, but scientists must have the final approval of people to work directly with them and to be involved with their research. If there are personal idiosyncrasies that a scientist cannot tolerate (race, religion, gender, or physical handicaps are *not* idiosyncrasies), it is best to be honest about it and eliminate those who evince those traits. Personal habits, manner of dress, and general social bearing can affect the way people feel about those with whom they are closely associated day in and day out, particularly under stressful conditions.

Of course, not all scientists are stupid or uncaring; in fact, there is a tendency on the part of some to go too far in the other direction. In giving credit for published work, for example, some researchers include the names of everyone in the laboratory who was remotely associated with the work leading to the publication. That kind of indiscriminate recognition downgrades the honor and destroys a most effective means of rewarding genuine contributions to laboratory research.

Balancing the Needs of the Laboratory and the Needs of the Scientist

The best research is done in an atmosphere characterized by freedom, and one of the most complex managerial problems lies in determining the boundaries of that freedom. It is a rare laboratory that can permit researchers to do whatever they like without reference to the mandate of the institution. Most scientists join the research staff with a plan that is well articulated when they arrive, and in fact staff appointments are often made because the work a scientist is doing fills a need in the laboratory. This does not mean that a researcher may not deviate from the original plan, but it does establish that the scientist and the laboratory have compatible goals. In government and industrial laboratories, the question of freedom is more sensitive than it is in academic settings because the mission often is quite specific. However, good industrial laboratories recognize that superior researchers must have the opportunity to pursue original ideas in addition to fulfilling their obligations to contribute to the programs of the laboratory. Robert A. Maxwell, who heads the pharmacology department of the Wellcome Research Laboratories has said, "The product-oriented lab does not necessarily stifle basic research; it is merely a question of meshing interests. If a scientist finds an industrial lab that is interested in the same things he is, there is no problem. . . . I can spend 50 percent of my time on basic research."

It is important to most scientists that they have colleagues with whom they can discuss their work. When adding a new staff member, this must be a consideration. Someone whose work is not sufficiently related to that of others in the laboratory may become isolated or may demand an excessive amount of the director's time and attention. However, in some places there are other institutions nearby where researchers can find the congenial colleagues missing from their home laboratories. Communication among these scattered scientists can be fostered by exchanging seminars, holding joint meetings and workshops, or other collaborative activities.

Fruitful discussion with colleagues does not necessarily imply that all parties to the discussion must be in agreement. It may not always be desirable. Studies have shown that scientific groups that retain their vitality over long periods of time often consist of individuals who, even though they may like and respect each other, remain intellectually combative.[5]

The mathematician S. M. Ulam wrote in his memoirs,

> Mathematicians are also prone to disputes, and personal animosities between them are not unknown. . . . When I became chairman of the mathematics department at the University of Colorado, I noticed that the difficulties of administering N people was not really proportional to N

but to N^2. This became my first "administrative theorem!" With sixty professors there are roughly eighteen hundred pairs of professors. Out of that many pairs, it was not surprising that there were some whose members did not like one another.[6]

The wise director accepts, even welcomes, tension that develops out of disagreement on technical strategies, theoretical approaches, or interpretations of newly uncovered knowledge. Out of such tension often comes the kind of scientific excitement that stimulates creativity and opens up new avenues of experimental thinking. What may seem like discord or even hostility between or among colleagues may be the struggle for truth, the ultimate goal of science.

It is important to be able to distinguish between healthy ferment and destructive dissension, however. The ideal is to maintain an atmosphere in which disagreement and debate are encouraged, expression of original ideas welcomed, and freedom to pursue ideas protected, without allowing anarchy to set in.

REFERENCES

1. Bush, V. *Endless Horizons.* New York: Arno Press, 1975. p. 180.
2. Levy, A.W. The Footnote: Sweepstakes. *The Sciences* 17(5):20, 1977.
3. Raju, M.R., Phillips, J.A., and Harlow, F. (eds.) *Creativity in Science—A Symposium* (LA-10490-C). Los Alamos, NM: Los Alamos National Laboratory, 1985. p. 69–70.
4. Sindermann, C.J. *The Joy of Science: Excellence and Its Rewards.* New York: Plenum Press, 1985. p. 23–27.
5. Pelz, D.C. and Andrews, F.M. *Scientists in Organizations* (rev. ed.). Ann Arbor: University of Michigan Press, 1976.
6. Ulam, S.M. *Adventures of a Mathematician.* New York: Charles Scribner's Sons, 1976. p. 91.

Chapter 5

ORGANIZATION OF RESEARCH

Orden no es una presión que desde fuera se ejerce sobre la sociedad, sino un equilibrio que se suscita en su interior.
[Order is not a pressure which is imposed on society from without, but an equilibrium which is set up from within.]
— José Ortega y Gasset, *Mirabeau o el Politico*

It is sometimes said that science is a creative pursuit that may be irremediably crippled by designating channels of authority through which certain processes must travel. The source of this belief is a soundly based fear of an atmosphere in which slavish adherence to procedures is more honored than expeditious service or prompt decisions. That, of course, is regimentation.

Organization, in contrast, is a method of assuring orderliness in relations among people, smoothness in daily operations, and a free flow of ideas within a conceptual framework. An environment incorporating these elements can exist only when functions and responsibilities are understood. An organization chart may be necessary to indicate the levels of responsibility, but it should be clear that position on the chart is not to be confused with scientific stature. The leader of a research group has both administrative and scientific responsibilities, but it is not unusual to find members of the group whose scientific achievements equal, or are even greater than, those of the group leader. The organization chart should serve as a guide, a kind of road map, and not as grounds for indictment or a trap to confuse and impede the work of the laboratory staff.

SIZE AND ORGANIZATIONAL STRUCTURE

In large measure, the size of a research laboratory determines its organizational structure. Some believe that certain psychological factors tend to diminish the effectiveness of very large research organizations; that they are likely to overspecialize in certain areas, become administratively ponderous, and lack flexibility to change course when new developments occur. It is true that certain laboratories may exhibit these trends as they grow larger, but certainly not all do. One rarely hears it said that Bell Laboratories or IBM research have grown rigid or have failed to keep in the forefront of research and development. The danger is not in growing, but in the failure to manage growth so that the fundamental nature of research remains in focus; new scientific ideas come from individual scientists and not from "manpower." When that focus is lost, something happens to the creative potential of the research laboratory. Provided that does not occur, a large laboratory can be more stimulating in some ways than a small one. Scientific creativity tends to flourish in the presence of a judicious mixture of disciplines, which is possible only in reasonably large organizations.

In universities, disciplines are usually divided into departments, each headed by a chairperson with clearly defined responsibilities. Rockefeller University is an exception; it has no departments but is organized into about sixty laboratories, each dominated by one or two senior scientists. Younger scientists have less independence than they might have in other universities, but Rockefeller has always attracted a large number of distinguished scientific scholars, and, as one young researcher commented, "Plato was willing to make a hell of a sacrifice to sit at Socrates' feet."

Government laboratories that are administered by universities tend to follow the university departmental pattern, at least in part. Brookhaven National Laboratory, for example, is managed by Associated Universities, Inc. (AUI), a consortium of nine universities. Brookhaven is divided into research departments, each headed by a chairperson, but it has also established groupings labeled as "divisions," each headed by a director. Oak Ridge National Laboratory, which in the past has been managed by Monsanto, and Union Carbide, and is currently managed by the Martin Marietta Corporation, is also separated into research divisions headed by directors. It is impossible to tell the difference between a Brookhaven department and an Oak Ridge division. Both laboratories are very large and the division/department design creates manageable components. At both places there is a great deal of collaborative work between scientists in different departments or divisions. Scientists have easy access to colleagues in a wide range of disciplines and frequently form teams, with whom they attack difficult problems and with whom they hold discussions to keep themselves professionally sharp. Brilliant scientists need people to talk with who are equally bright; ongoing interactions between scholars of equivalent caliber in different disciplines are more likely to occur in larger research organizations and frequently produce fruitful new research ideas.

The lines between basic and applied research are likely to be blurred in industrial laboratories, where the ultimate objective is usually product development. Yet many product-oriented laboratories encourage their research staff to devote time to basic research, even though the body of knowledge in almost every field of technology has reached such a scale that no industry can expect to be entirely self-sufficient in generating all of its technology, let alone the science on which its technology rests. Creative leaders in industrial laboratories recognize that the quality and tone of a development organization are importantly influenced by the presence of good basic research activity. They have also found that a basic research group provides management with a source of competence for technical appraisal in looking into the future. Frequently, research groupings in industrial laboratories are built around specific projects, and the project leader assumes technical and administrative responsibilities likely to be more extensive than those of a university departmental chairperson or even a group leader in a basic research laboratory.

GROWTH OF NEW LINES OF RESEARCH

No active scientific grouping or arrangement continues for very long without change. Research, by its very nature, tends to grow; healthy research activity sends out tendrils that seek new ground and new cultivation. Managing the growth of a research organization involves deciding which of the offshoots are to be fostered and which allowed to wither. When it becomes obvious that one particular line of research has become very strong and is tending to dominate the department or group in which it is located, the director must make a judgment about its value and decide what to do about it.

There is no foolproof formula for deciding which are good ideas, which research will turn out to be the key that opens the door to important discoveries. Usual methods of evaluation—consulting experts, searching current literature, asking advice of senior scientists on staff—may or may not result in correct decisions. Roy Rowan believes that intuition, or "the Eureka factor," as he calls it, is as good or better than any other tool used in decision-making, and he has assembled an impressive number of testimonials in its favor in his book, *The Intuitive Manager*. Rowan and the famous executives he quotes have great faith in the hunch.[1]

When a decision is made, on whatever grounds, to transfer a research program from a group and set it up as an independent unit, the transition ought to take place with the least possible interruption. Such moves will also bring about changes in allocation of facilities, personnel, budget, and services. Technical and administrative service departments should be reviewed to ensure that they can handle increased workloads that the changes may impose.

Universities have found that setting up research centers or institutes outside of departmental structures is one way to handle a program that may have become too large or too dominant for the department. Research centers and institutes provide vehicles not only for dissolving departmental competition, but also for collecting a range of expertise to concentrate on a "hot" problem. They create a focus for funding agencies interested in particular areas. Indeed, a center may be the ideal solution for an activity that, for whatever reason, has become awkward to administer within a department, so long as the program merits continued support and cultivation. Setting up an independent unit is one way to coalesce the efforts of a multidisciplined faculty to focus on a central problem, permitting collaborative research with others in overlapping or peripheral fields without the restriction of departmental separation of people, funds, and facilities.

Research centers can also establish a research reputation on their own, separate from their parent academic departments. The university department may have become associated in the public mind more with education than research.

Yet there are certain drawbacks to establishing research centers or institutes that university administrators should be aware of: They may create problems with maintaining a balance in the particular research area. What's more, academic chairs and research-center chiefs may become enmeshed in conflicts over scarce resources and status. And there is always the possibility that the center will overshadow the educational function of the institution.

SCIENTIFIC FREEDOM AND ADMINISTRATIVE RESPONSIBILITY

Scientific administrators have a responsibility to see that their laboratories uphold the greatest value and oldest tradition of science: to tell the truth. This means that close attention must be given to the day-to-day laboratory operations. Stewart Blake, a researcher and director of research and development programs, commented at a panel on management for creativity: "I feel very strongly that as much attention has to be given to the day-to-day procedural operations as has to be given to the broader questions such as the determination of the goals."[2]

Attention to daily operations does not necessarily mean telling scientists what to do: It surely means seeing to it that whatever they do is done in a proper way. Data must be scrupulously recorded in laboratory notebooks and retained. Every experiment must include provisions for the protection of personnel, animals, and the environment. The laboratory director may be called upon to assist in data interpretation or to suggest qualified experts who should be consulted. The scientific administrator's responsibility also includes seeing to it manuscripts are rigorously reviewed to ensure that publications emanating from the laboratory meet the highest scientific standards. And, no matter what committees or which outside experts or journal editors have reviewed a manuscript, responsible senior scientists or laboratory heads should make certain that *all* names appearing as authors or coauthors on the paper are those who actually participated in the research and were active in, or fully knowledgeable about, the collection and interpretation of the data.

No laboratory director can be present in every research module to observe every activity. In many laboratories, the director is not sufficiently knowledgeable about all areas of research going on to review every manuscript personally. Even if one's knowledge were very extensive, surely there would not be time enough to give each piece of work careful consideration. Obviously, many responsibilities must be delegated to others who understand the purposes of an institution and its management philosophy.

Usually scientists are chary of accepting responsibility for their colleagues, especially those who are considered independent research workers. Traditionally, those who achieve that level of recognition are expected to stand or fall on their own work. In those comparatively rare cases when they do fall, unfortunately, they nearly always bring down one or two, or sadly, many colleagues with them, ultimately tarnishing not only their own names and those closely associated with them, but their home institutions' as well.

When John R. Darsee, a young researcher working at the Emory University School of Medicine and Harvard University Medical School, was found to have fabricated much of his research data, investigation revealed that he had collaborated with forty-seven other scientists at Emory and Harvard, and altogether they had coauthored eighteen scientific papers, eighty-eight abstracts, and three book chapters based on the fraudulent data. Although none of Darsee's collaborators was charged with fraud— and no evidence was presented to show that they were directly involved in the fabrication—their participation in perpetuating fraudulent information was evidence of grave errors in judgment and shocking carelessness. They had become parties to a serious violation of scientific veracity.

On April 22, 1986, *The New York Times* reported on a study of the Darsee case conducted by two government investigators. The *Times* quoted the study's summary of the collaborators' misdeeds:

> Publishing papers with a very large number of obvious errors and discrepancies; publishing data that seemed highly implausible; failing to preserve original data so that others could review them; putting their names on scientific papers on which they had done little work; and republishing old data in a new form without indicating they had done so.

Fearing libel suits, no publisher could be found for the study for some time. In January 1987, however, the results of the Darsee investigation were published in *Nature*.[3] How easy it would have been to prevent the whole affair by the exercise of simple internal controls! There were innumerable opportunities for senior scientists and administrators to have questioned the data, but, apparently, no one did. The review mechanism obviously failed; the fact that a scientist had not preserved his original data would surely have come to light if only one of his forty-seven collaborators had questioned the data that one Emory scientist later described as "so fantastic it should have been questioned by anyone who read it." All the errors and misstatements escaped the attention of the coauthors, journal editors, and experts who reviewed Darsee's papers. The offense (or error in judgment) of putting one's name on an article on which one has done little work is not uncommon. The practice has created problems in many laboratories and has, in some cases, been a source of scientific embarrassment. Nearly always, the blame falls on the career demands of scientists to "publish or perish." Inadequate grounding in scientific ethics causes the problem, but management must take some of the responsibility as well. While research organizations must offer their scientists great freedom, such freedom is not a natural right. It is earned, based on distinguished performance over time.

The research administrator, together with those who have assumed delegated responsibilities, must respect the need for their colleagues to work in their own way, to bring their own talents to fruition. But, at the same time, they must learn to interact with their colleagues in such a manner that their own role is fully understood and respected. The judgment of the individual scientist is not sacrosanct. When there is reason to believe that scientific standards, or those of ethical conduct, are being infringed, the matter must be investigated fully, impartially, and promptly.

The Darsee case suggests that improvements in the review system are urgently needed, not only in research laboratories but at publishing organizations as well. In their defense, journal editors complain of being overloaded with reports from ambitious scientists eager to increase their publication records. They claim that it is almost impossible to tell whether old data are being published in a new form. Editors point out that they are dependent upon the advice of the experts to whom they send papers for review. Obviously, if the referee system fails, as happened in the Darsee case, something may be wrong with the system—or with the standards for selecting referees.

A SENSE OF PURPOSE AND COMMUNITY

The key ingredients for a successful research organization are people, purpose, and resources. Ideally, to create a top research group, you would assemble a brilliant research staff supported by gifted technical and administrative assistants, located in a salubrious climate, with sufficient resources to provide generously for facilities, equipment, and salaries—plus a noble and inspiring purpose. No such place exists, of course, but the creative research administrator never ceases to strive to achieve it.

There is no end to the struggle to improve facilities that make research possible: work spaces, instruments, equipment, libraries, and assistance in making use of them. The search for scientists of the highest intellect and greatest creativity is continuous. The endeavor to generate and maintain the will to accomplishment and pride in the institution is never completed. That will and that pride, rooted in a thorough understanding of the highest purposes of each institution, must be cultivated in everyone—not only the research staff, but also people in the shops, those ordering the supplies, typing letters, and washing glassware. Every staff member needs to have a feeling of participation. It is surprising how much people can accomplish, even under highly adverse circumstances, when they feel that they are part of a worthy project and understand what is happening.

The principal purpose of organization in a research environment is to enhance the free flow of creative thought by clearing out the underbrush of administrative and technical "housekeeping," so that the scientific talent and energy are fully directed to research. The director, to keep perspective, should remember that "When you're up to your neck in alligators, it's easy to forget that you started out to drain the swamp."

ADVISORY COMMITTEES AND CONSULTANTS

No matter how wise, brilliant, or accomplished your staff may be, and no matter how good the chief administrator is, a first-class research laboratory must depend on advice, assistance, and guidance of others. The greatest of all wisdom, perhaps, is to know what one does not know and to know what to do about it. The best way for laboratory leaders to augment their knowledge and expertise is by selecting qualified advisers who act as consultants both inside and outside the laboratory.

Committees have been the subject of almost as many jokes as mothers-in-law: "A camel is a horse built by a committee"; or "The best way to keep something from happening is to appoint a committee to do it"; and so on. Within these witticisms lies a kernel of truth. But to be fair, it can be pointed out that although a camel is not a horse, it is useful in places where a horse would not do.

Management of a scientific institution can be divided into two general areas:

1. Policy exploration and formation
2. Policy execution

Committees can contribute a wide range of thinking in the exploration and formation of policy. Their role in policy execution, however, should be limited to a few specific areas, mostly involving laboratory safety and ethical research practices, where committees are effective executors of policies.

The ideal way to resolve some problems is by allowing them to defuse while a committee debates the issue. And sometimes appointing a committee gives the institutional head time to ascertain the facts and arrive at an independent conclusion while the committee confers. Occasionally, the mere act of referring a matter to a committee serves as an immediate solution by demonstrating to the interested parties that the matter is being taken seriously. Committees are, however, indispensable in the review

and evaluation of research programs. Properly used, committees can contribute a great deal to formulation of policies and planning of technical
and administrative services and in matters of general welfare.

External Committees

Outside committees are most useful in decisions on scientific matters,
particularly in evaluating research programs, and sometimes in the exploration and formation of policy. They provide a body of opinion representing several areas of knowledge and expertise. Their combined wisdom
and experience in different areas of specialization bring deeper and
broader evaluative judgment to the research under review than any one
person could offer, no matter how knowledgeable. Outside advisers also
give the head of the laboratory important backing for sensitive decisions.

A negative evaluation of a research program by an outside committee
composed of highly respected scientists has more credibility than one
given by the research director or even by an internal committee. Also, it is
easier for the research staff involved to accept a poor review from that
source. A good evaluation from an outside body is enhanced by the
knowledge that the quality of the work is recognized beyond the walls of
the laboratory where it is done.

When there is a decision to be made about cutting out a program, the
opinion of a group of widely known, outstanding experts in the field
provides powerful support for the final judgment, whatever it is.

In selecting external advisers, outstanding experts are exactly what
the laboratory should aim for. Established scientists with impeccable credentials respected by the scientific community are not easily recruited as
staff members, particularly if they have a long-standing affiliation with
other reputable institutions. Still, they are often willing to serve as research advisers. Their association with the laboratory through committee
visits strengthens the prestige of an institution, and if their duties are
conscientiously performed, its research program will be strengthened.

It is only reasonable to assume that when top people in a field are
invited to serve on a committee, all will not accept. People of such qualifications are invited to do so many things they must struggle to set aside
time for the very research on which their prestige is built. But it is a
mistake to rule out anyone on the basis that they will decline. If you really
believe someone to be the best possible choice, go ahead and issue the
invitation. Let the decision be made by the one who is invited.

In some highly competitive areas of scientific research, where disagreement on theoretical approaches has divided the scientists into
camps, selecting objective outside program reviewers becomes difficult—
but it is essential. Choosing experts known to favor a particular view casts
doubt upon their recommendations; yet, someone seen as a rival or be-
longing to an opposite camp will be equally suspect. It is particularly
important to keep in mind that final decisions are always in the hands of

the institutional head. The consultant's role is to bring the widest possible range of informed judgment to bear on each matter.

Invitations to serve as an adviser in any capacity must state clearly what is expected: how often advisers are to visit the lab or conduct reviews of research programs; what reports and recommendations they are to submit and in what form; if their advice will be sought on specific matters between visits; and under what circumstances additional visits may be necessary.

Commitments on the part of the laboratory should also be clearly spelled out: what materials will be supplied to help advisers prepare for program reviews; remuneration including honoraria, travel, and living expenses; and any other special considerations. It is wise to be quite specific about reimbursable items in order to avoid misunderstandings, particularly in enumerating special incentives, such as travel and living expenses for spouses or a rental automobile during the visit.

The number of visits anticipated should take into account the time constraints of advisers. By all means, refrain from unnecessary demands or from expecting visiting experts to deal with matters that can best be handled internally. Making unnecessary demands on their time is a good way to lose valuable advisers.

Advisory committees should be appointed for a period of at least three years, longer if possible. It will take some time to become acquainted with the research staff and the work of the laboratory. The longer the acquaintance, the sounder the judgments are likely to be.

One of the most difficult problems in dealing with outside advisory committees lies in scheduling visits. Finding two or three days together when several eminent scientists will all be free to come for a visit is not easy. The schedule must be arranged as far in advance as possible (a year is not too far ahead) with tactful reminders sent from time to time, preferably enclosed with informative material. Yet, even after the most meticulous planning, it is not unusual to find that the schedule must be shifted in order to have the whole committee present. That is the price to be paid for peopling your committee with top scientists and it is usually worth it.

External Consultants

In selecting individual consultants, qualifications are the same as for advisory committees—outstanding knowledge and the respect of colleagues in the research area under study. Scientists have unwritten understandings with their colleagues that they will review each other's papers, respond to one another's questions, or react to ideas on a reciprocal basis without payment. Many researchers keep in touch with their mentors long after completing their education or thesis research, asking them for guidance and opinions on research matters, manuscripts, grant proposals, and so on. It is easy for a new administrator, who has habitually called upon a colleague or mentor for assistance and advice to continue the practice, overlooking the fact that their relative positions have changed. When one becomes an institutional administrator with funds at one's disposal, the expectation of reimbursement for advice and consultation automatically arises among colleagues who previously provided such services gratis. The new laboratory head should be aware of this and should formalize agreements for services such as reading and commenting on manuscripts, reviewing and advising on research programs, and consulting on scientific, administrative, or organizational matters.

Industrial laboratories, national laboratories, and some academic institutions have standard contracts for consultants with the boiler plate in place. Consultants may be appointed for a specific purpose or a certain period of time, or, if they are to serve on a continuing basis, the contract may be renewable annually.

In a small or new laboratory where there is no standard consulting agreement the temptation is great to negotiate with consultants verbally, but it is not advisable. A simple letter of agreement serves the purpose of spelling out commitments on both sides, and clarifies expectations of each. Even if no payment is involved, it is a good idea to put *that* in writing, too, and include some indication of appreciation for the service, such as, "I am sorry we cannot pay you an honorarium for this service, but we will be glad to return the favor"; or "Our budget does not allow us to offer you an honorarium, but we will wine and dine you and take you sailing on the lake on Thursday afternoon" (or for a visit to the nearby pueblo, or hiking in the mountains, or to the Metropolitan Opera performance of *La Bohème* in the evening).

The matter of confidentiality can be a sensitive issue in relation to individual consultants or with any program reviewer, especially in fast-moving, highly competitive fields. Confidentiality tends to be of much greater significance in industrial and government laboratories where security measures have long been standard. Academic institutions are inclined to favor as much openness as possible in order to ensure that scientific information is freely shared. The philosophy and training in academic settings includes the belief that everyone's work builds upon the work of others and the goal is not (*as a rule*) immediate development of a product. But even in basic-research laboratories, there is always competition for early publication and scientists who are in a race with others in "hot" fields may prefer not to discuss the details of current activity with anyone. Up to a point, the laboratory director must respect each scientist's right to remain circumspect and consultants or reviewers must understand this principle. But a cloak of secrecy should not be permitted as an excuse for lack of progress, or simply for doing nothing.

There are several common failings or blind spots regarding outside advisers. One of the major errors is to forget about them between visits. On a routine basis, advisory committees and individual consultants should receive publications, reports, and any other documents relating to the programs of their concern. If the laboratory circulates a newsletter or internal bulletin, such publications should be available to advisers. It is also a good idea to invite outside consultants or committee members to special events such as conferences, important lectures, significant ceremonies, or major social events. If advisers feel that they are part of the institutional family, they are more likely to identify with its struggles to advance excellence and to be more generous with their time and talents in the fulfillment of their advisory duties.

Another common mistake is failing to provide the necessary information for a review visit sufficiently in advance for advisers to be informed about the current status of programs to be reviewed. It is not unusual for visiting reviewers to be confronted with bulging packages of material on the day they arrive when there is not enough time to read, much less absorb, so much material.

The most baffling of all, however, is the practice of appointing committees or consultants and then ignoring them. Some institutions appoint an outstanding panel of consultants in various fields, decorate their publications and letterheads with illustrious names, point with pride to their affiliation with the institution, mention their names in grant applications, and never call upon them for advice. Or they may be consulted rarely and in such a disorganized manner that their talents and the honoraria paid are wasted. If you go to the trouble to recruit good advisers, by all means, use them.

Internal Consultants and Committees

In large research organizations such as national laboratories, qualified advisers in numerous fields may be found within your home institution. If the finest expertise can be found in one's own backyard, it should not be overlooked. Internal consultants offer the advantages of convenience, informality, and relatively low cost—at the very least, there is no travel expense. Consultants within the same institution are most likely found in fields that are peripheral to the research activity concerned—chemists to advise biologists, or engineers as consultants to physicists.

The major drawback of internal consultants is that because of the informality and proximity, there may be social bonds between researchers and advisers that may weaken objectivity. There is a greater likelihood that a consultant will fall into the trap of being co-opted by one group with a special theoretical bias. It is easier to find charming, articulate, and compatible individuals more credible than contentious or abrasive ones, but personality is not always a clue to soundness of judgment or brilliance of research ideas.

Internal laboratory committees are indispensable in exploring and resolving issues that relate to general management policies of your institution, such as personnel procedures, safety, technical and administrative services, and anything relating to the general welfare of the staff.

Scientists are not anxious to spend time in the committee room; consequently, this management tool should be used sparingly. The mission of a committee must be clearly articulated, and, unless it is a standing committee, it should be disbanded when its mission has been accomplished. If committee appointments are made with discrimination, if demands made on the group are reasonable, and if the recommendations submitted are treated with appropriate respect, being named to a committee becomes something of a distinction and staff members are more willing to accept the additional responsibility.

At times there is a temptation for laboratory heads to appoint members to committees whose views are most likely to coincide with their own, but surely this is a mistake. A body composed of people with differing opinions will contribute a wider range of ideas and thought than will a homogenous one. After free expression of conflicting views, moreover, those who did not prevail will be somewhat mollified by having spoken their piece. ("At least I let them know what I think!") And the very fact that representatives of all sides of an issue were consulted and given the opportunity to express an opinion goes a long way to ensure compliance once a ruling or decision has been reached.

Internal standing committees are essential in some cases (such as those concerned with safety, human and animal subjects, and the like). Membership on standing committees is determined largely by professional qualification, but within the confines of that limitation the wise administrator will still try to assemble a balanced group of scientific interests, ethical views, and personality.

Ad hoc committees may be appointed to make recommendations on matters of general concern. These will often deal with administrative and technical services and facilities or with questions of personnel policies. When funds for general-use equipment are limited, as they almost always are, representatives from user groups should be consulted about priorities that guide such purchases. They will identify the most needed item to be ordered first, and those who favor another item for first priority will understand how the determination was made. In a large laboratory, or even a modest one, the location of a central facility, such as the stockroom or cafeteria, and the hours they are open are important to the entire staff, not just scientists. Before making any major change in a central service, a committee composed of representatives of all those affected should be consulted.

Staff meetings, attended by senior scientists, are held in some laboratories. In a lab where all research is focused on one area, the senior staff may serve as a research advisory committee, providing a forum for discussion of research matters that concern the entire program. In larger places, where research interests are varied, regular staff meetings provide an opportunity to convey relevant information to the scientific staff, or for individuals to bring up questions or identify matters that are of concern.

These discussions may pinpoint general problems that committees can be helpful in resolving.

Laboratory heads should always appoint committees in which they place full confidence and, except in rare cases, should always follow the committees' recommendations, even when they do not agree with the advice of the committees. Otherwise, cynicism sets in.

Occasionally there is some excellent reason for disagreement with a committee decision. The director might have certain information that the committee could not be made privy to; in this case, the fact that full information was not available to the committee should be stated when a ruling in disagreement with the committee's recommendation is made. In highly controversial matters, when there is heated debate in the committee room, the lab head may choose to differ with the final recommendation on the basis that it was not a clear-cut decision, with the acknowledgement that "I am in the position of casting the deciding vote."

REFERENCES

1. Rowan, R. *The Intuitive Manager.* Boston: Little, Brown and Co., 1986.
2. Raju, M.R., Phillips, J.A., and Harlow F. (eds.) *Creativity in Science—A Symposium* (LA–10490–C). Los Alamos, NM: Los Alamos National Laboratory, 1985. p. 102.
3. Stewart, W.W., and Feder, N. The Integrity of the Scientific Literature. *Nature* 325 (6101): 207–214, 1987.

Chapter 6

COMMUNICATION

He had forty-two boxes, all carefully packed,
With his name painted clearly on each:
But, since he omitted to mention the fact,
They were all left behind on the beach.
                    —Lewis Carroll, *Hunting of the Snark*

Science does not stand still. And it does not stay in one place. Something that is discovered in a laboratory in Bloomington, Indiana today may affect the work done in Nagoya, Japan next week. Dissemination of research information is second in importance only to the performance of the research itself. And it is not enough to communicate with other scientists—the public now demands to know the latest developments.

Public demand arises from a mixture of admiration and fear. Just as the Industrial Revolution brought us wealth and filth, modern science has brought us enlightenment and terror. Even as we enjoy the comforts ("Better things for better living," as one chemical company used to advertise) of advanced medical knowledge, ease of travel, and other blessings of research, we are horrified by its achievements in the art of destruction. More subtle, but no less frightening, is the growing awareness that a by-product of the scientific marvels we admire is the slow pollution of the atmosphere of our planet that may erode its capacity to support human life. People have become suspicious of the scientific community, ready to question and, at times, oppose the installation of research facilities and to mount popular demonstrations against certain kinds of research. Dissemination of scientific information to the public and to the scientific community falls into the category of "External Communication."

EXTERNAL COMMUNICATION: SCIENTIFIC CONFERENCES

"Pro-gress or Con-gress! You can't have both. A scientist must choose," says Raymond Latarjet, who for many years was head of the Institut du Radium of the Curie Foundation in Paris. Latarjet, besieged with invitations to speak at or participate in conferences, decided to curtail travel to meetings in order to devote more time to research and the administration of his institute.

Other scientists consider gatherings worth the sacrifice and some spend as much or more time at meetings as they spend at the laboratory.

Professor J. H. Burn, who directed pharmacological research at Oxford University from 1937 until 1959, was frequently persuaded to attend international and national meetings, much to the annoyance of the researchers at his laboratory who depended heavily on his guidance and were somewhat resentful when he was not around to discuss their problems. They would say, "The lab fiddles, whilst Burn roams."

There is, of course, a middle ground. It is true that the most effective way to communicate research information, as well as practically anything else, is in person-to-person conversation. And the second best is to talk before groups with the same interests at conferences, seminars, and workshops. As it turns out, a great deal of information exchange at conferences takes place not in the lecture hall, but person to person in corridors or over dinner tables. Acknowledging this fact, some scientists are invited to be "participants" at meetings, where their presence is the only contribution they are expected to make. Just being present to talk with others, particularly less-advanced colleagues, adds enough to the success of the meeting to make bringing them there worthwhile. And it is well known that a program, even one marked "tentative," dotted with a few illustrious names will attract early registrants who rush to attend any meeting that Dr. So-and-So may attend.

Researchers should be encouraged to attend appropriate gatherings where they are invited to speak about their work or to learn about the work of others. Funds for official travel should be included in every lab budget. The institution should also assist staff members in obtaining support for travel from outside sources, particularly for meetings held abroad or at great distances.

The laboratory should be active in organizing and hosting gatherings of scientists. Research personnel should also have opportunities to serve on committees of scientific societies concerned with the organization of conferences, workshops, and other meetings.

Scientific meetings serve many purposes. Annual meetings of professional societies are common occurrences, and an active research organization makes a point of holding research conferences on subjects of current or emerging interest. It is one way of getting the latest information on a "hot" topic. The trick, of course, is in picking the topic that will be hot when the meeting takes place, because planning a large, first-class scientific conference takes at least a year. Managing to do this on an annual basis is a challenge that few laboratories are willing to attempt. The Oak Ridge National Laboratory, from its beginning soon after World War II, established a pattern of holding annual information meetings in each of its approximately twenty divisions. Alvin Weinberg, who was director of the laboratory from 1955 to 1974, says of these meetings:

> We tried to combine at least part of the information flow function with a sense of periodic renewal through these meetings. They were an opportunity for each division to display its intellectual achievements, its intel-

lectual wares, to an advisory committee. But they were also a kind of minicelebration of the end of one year and the beginning of another.

And, aside from the information meeting, the biology division held a research conference every year, which attracted speakers, participants, and observers from all over the world.

The biology division was the driving force behind a series of Latin American conferences, organized to promote cooperation between scientists in the Americas and to bring Latin American science to the attention of laboratories in Europe and Asia. Alexander Hollaender, who instigated these meetings, called them "scientific windows on the world," where the host institutions observed others and were, in turn, observed by them, counterbalancing the effects of geographic or scientific isolation.

The topic of a research conference should deal with an area of concentration at the host institution and preferably one in which the laboratory has achieved some reputation for its contributions.

The Organizing Committee

Sometimes the topic is the impetus for the meeting; that is, the idea of a conference arises when several people, perhaps several research groups, are experiencing the need for enlightenment or feeling frustrated at their lack of progress. The hope is that by meeting and pooling their knowledge all will benefit. No matter how the topic is chosen, the organizing committee must be made up of researchers at the host institution who are actively working in the field. They are likely to be those most highly motivated by their interest in the subject, and also the ones who know the leading scientists who should be invited to participate. Such knowledge goes a long way in designing a program rich in content and attractive to others with similar research interests.

When the organizing committee is appointed, a conference manager should also be named to handle correspondence, record-keeping, and registration. The manager will probably be someone from the administrative staff, but that person should attend every committee meeting from the beginning, since the more managers know about planning and negotiations, the better the job they will do.

Setting the Date

Many traps lie in wait for the conference planner, but among the more dangerous is setting the date. Nothing will ensure failure more certainly than selecting a date that conflicts with that of an already announced meeting of importance to people working in the field of the conference subject. Careful research must be conducted to ensure that no scientific society or other significant research gathering has been scheduled for the chosen date. Then lose no time in disseminating far and wide the date of the meeting as soon as possible to prevent others from falling into the trap you have just avoided.

Climate and geography, too, should be taken into account when considering the date. If you are in a region where sudden snowstorms may make travel difficult, avoid a winter meeting unless you hope to attract a skiing crowd. If you are in the Sunbelt, a meeting during the cold months has great appeal. Although the area's recreational opportunities may be a special incentive for registrants, they can also be a major distraction. An isolated locale with few diversions (at least during the daytime) usually results in higher attendance at meeting sessions.

Making Up the Program

The selection of speakers at any scientific meeting is a highly sensitive issue. In his gently satirical, delightful article "Where the Elite Meet," David Owen Robinson writes:

> In order to attract all the *right* people, it may be necessary for the organizer to have a few great names already signed up before the recruitment of other participants begins. Thus, even the lowliest assistant professor can assemble a remarkable array of talent: "Dr. Einstein? We were wondering if you would be interested in joining our symposium on the biological determinants of artistic expression. We think Michelangelo is going to be there and we've already got Dr. Schweitzer and the Wright brothers coming." . . . Who could refuse?[1]

Although one rarely succeeds in snagging all the most desirable speakers, an astute organizer will see to it that one or two, at least, are well known and highly respected in the field. It is most unwise, however, to list names in a published program until they have formally (in writing) accepted your invitation. An early listing of speakers in a program clearly marked "tentative" may include those who have been invited and promised to think it over and give their response later. It might be possible to get away with misleading potential attendees once, but even once may damage your laboratory's credibility and may jeopardize future efforts to conduct successful conferences.

While every field has its "stars" whose participation in a conference assures that it will be taken seriously, every field also has its up-and-coming young people who are not yet well known. It takes some research to identify them, but it is worth the effort. The stars of tomorrow are just as likely, maybe more so, to make impressive presentations at a conference as are the stars of yesterday and today. Suggestions should be sought, not only from staff researchers, but informally (and confidentially) from colleagues and friends at other laboratories. In that way, a young postdoctoral fellow or a brilliant graduate student may be discovered who is doing significant work related to the conference topic. Asking colleagues to nominate speakers or participants for a meeting can create misunderstandings and arouse false expectations, however, so it must be emphasized that suggestions are being requested from a wide range of scientists and not everyone whose name is put forward can be invited to be on the program.

Aside from inviting a certain number of speakers, the committee may also want to invite session chairs and discussion leaders; therefore, it will be helpful to collect many names of people doing good work in the field, even if they cannot all be main speakers. Also, everyone whose name is suggested should be placed on the list to receive the announcement of the meeting. That list can serve as the nucleus of your major mailing list, which must be developed as soon as possible. In the long run, the quality and stature of speakers will determine the overall success of the conference.

Only after the tentative program has been completed can the planning begin. For organization of large conferences, professionals may need to be called in. The local committee then makes up the program, gets the speakers, and decides on the location; the other details are handled by the pros. There are some very good ones with a reputation for being able to handle everything, including a fire at the hotel, but it may take some research to find the right organization for your meeting. If you are starting out with no leads from societies or groups that have had experience in dealing with professional conference managers, a good place to begin is by writing to Convention Liaison Council, 1575 I Street, NW, Washington, DC 20005. Ask for a listing of convention managers or meeting planners in your state or in your area. Some other organizations that might provide helpful information are:

Meeting Planners International
3719 Roosevelt Boulevard
Middletown, OH 45044

Professional Convention Management Association
2027 First Avenue North
Birmingham, AL 35203
(Primarily plans medical meetings)

Council of Engineering and Scientific Society Executives
2000 Florida Avenue, NW
Washington, DC 20009

If an exhibit is an integral part of the meeting, a good organization to consult is:

National Association of Exposition Managers
P.O. Box 377
Aurora, OH 44202

Conference Site

The meeting place must be selected as soon as possible; no announcements or promotional materials can be prepared until it has been confirmed. If it is to be held outside your institution, select a place that is at least within commuting distance of the laboratory. Speakers and others who attend may want to visit your laboratory and see, at first hand, the work going on.

Hotels and motels have found conferences a good way to fill up their rooms, particularly in the off-season, and many of them offer excellent conference services, meeting rooms with comfortable seats and audiovisual equipment, travel assistance for guests, and advice on recreation plans. Nothing should be assumed from hearsay regarding the quality and type of facilities offered by an establishment, except possibly in the case of famous big-city hotels. A representative of the conference committee must visit the proposed sites to assure that the facilities are adequate for the size and type of meeting planned, and to discuss the details with the hotel's conference manager.

Conference sites must be booked well in advance, as soon as the date for the meeting has been set and a determination made concerning possible alternate dates. Some flexibility in dates may be helpful in negotiations concerning costs and services at potential sites.

Special room rates can always be negotiated for guest accommodations, and the rates agreed upon should be obtained *in writing* before the information is passed on to conference registrants. The printed invitation usually includes information about the site and gives the range of room costs; registrants then arrange accommodations directly with the hotel. The committee may arrange for rooms for speakers and others whose expenses are to be paid by the conference committee, and they are given the best available accommodations.

Hotels and motels sometimes provide free meeting rooms for conferences if the meeting brings in a large number of guests; others charge for the space, regardless of guest rooms occupied. In some meeting rooms, audiovisual equipment has been permanently installed; in others the hotel will agree to have it brought in; in either case, the equipment should be tested before the meeting starts, and there should be an understanding as to whether the hotel or the committee arranges for a technician to operate it. It is sometimes better to have a technician who is familiar with the subject matter. Just be sure the technician belongs to the appropriate union!

On preliminary visits to inspect potential sites, ask about specific services hotel employees may be expected to provide, such as relaying telephone messages during meeting sessions, providing meals outside of regular hours, room service, and so on. Also ask about child-care facilities, special dietary meals, wheelchair access to guest and meeting rooms, and arrangements for coffee breaks.

If there is to be a banquet or reception during the meeting, menu choices and cost estimates should be discussed. Banquets are sometimes included in the registration fee, but it is usually better to make them optional and permit attendees to subscribe separately for themselves and nonregistered guests, if they wish. If a banquet is included in the registration fee, it will still be necessary to establish a per-person cost for those wanting to invite husbands, wives, or friends not attending the meeting.

Conferences of as many as three days in length should include on the program some kind of social activity or outing to break up the sessions, preferably on the afternoon of the second day. The staff at the conference site can usually help in making suggestions and arrangements for these outings—bus transportation to a marina or ski lodge, for example. If the site is in a city, on-site staff can usually furnish information about the availability of tickets to theaters, the opera, or concerts, or may make arrangements for special tours to museums or historic sites.

After a site has been approved by the committee (or its representative), details of arrangements can be worked out between the hotel or motel personnel and the conference manager. All cost quotations must be in writing, and in cases where a firm figure cannot be given, the basis for the cost should be stated. For example: "Two coffee breaks per day for 2½ days for approximately 200 people @ $1.00 per person; $2 \times 2\frac{1}{2} \times 200 \times \$1.00 = \$1,000$."

Quotations on coffee breaks and any other item involving provision of food and drink, such as cocktail parties, banquets, and so on, should state exactly what is to be served. Will there be tea at coffee breaks? Decaffeinated coffee? Hot chocolate? Sweet rolls?

The conference manager will handle the details but any major deviation from agreed-upon costs or services must be approved by the organizing committee. When advance registrations begin to arrive, the conference site and the manager must maintain very close communication; hence, an item in the budget should cover travel costs for the conference manager between the site and the laboratory, as well as other trips to printers, caterers, and so forth.

Budget

A meeting that depends upon outside sponsorship usually requires more planning and organizing time than one entirely funded by registration fees. But it is always good to allow plenty of time. As a rule, early planning should begin no later than a year before a sizable conference is to be held.

The most promising sources of support for conferences are government grants, foundations, corporations, or individuals who have an interest in the subject. Both the National Institutes of Health and the National Science Foundation make grants to support research conferences. Pharmaceutical companies are interested in research that may lead to new

drugs or development of medical devices. There is a wide range of companies that have an interest in chemical research: cosmetics, household-
cleaning solution manufacturers, the food industry, and even clothing
and automobile manufacturers. There are few social-science topics that
do not in some way impinge upon the world of marketing, and the world's
vast communications industry depends upon engineering, physics—perhaps all the sciences.

Success in recruiting sponsors depends heavily upon having at least
one person in the host institution with the energy, imagination, and ability to convince an organization or an individual that the knowledge to be
gained from the meeting will have relevance for the products or programs
of the potential sponsor.

An important point to remember about conference planning is that
some outlay of funds may be required before any fees or other monies are
collected—printing costs, deposits to hotels, caterers, and so on. The
laboratory must have a source for advancing the necessary funds or must
have established an excellent credit record in the community where the
conference is to be held.

It is possible, at least theoretically, to make a profit on a conference.
Kenneth A. Davis of the Greater Framingham Mental Health Association
in Massachusetts says:

> A one-day conference can easily net in excess of $1,000 while a two- or
> three-day conference can realize a profit of more than $5,000. The exact
> financial outcome obviously depends on many variables, such as format,
> size, location, speakers, and advertising; however, significant amounts
> can be guaranteed through adequate planning.[2]

An examination of sample budgets given by Davis, however, suggests
that a large proportion of conference expenses must be absorbed from
other funds, and that speakers live very nearby and earn very modest fees.
Presumably, he has based his estimates on the Framingham area, which
may not be applicable to all locales. The article contains a great deal of
useful information, however, including a conference-planning schedule
beginning a year in advance of the meeting. Social-science conference
planners will find it especially helpful. It has some worthwhile material
and ideas for any conference planner.

The skeleton budget, shown on page 84, is given without cost figures,
which will vary according to time, place, and circumstances. This guide
indicates the kinds of costs that must be taken into account.

Even a tentative budget cannot be made up until the site has been
selected. The costs of space and facilities serve as a basis for estimating
others. Some items, such as speakers' and chairpersons' fees or honoraria,
are entirely up to the organizing committee, which must determine an
amount that is appropriate and sufficiently attractive but not so large as to
force a shortening of the program or cutbacks in other important items.

Skeleton Budget

EXPENSE

Facilities:
> Meeting room(s)
> Audiovisual equipment rental
> Audiovisual technician(s)

Promotion and Advertising:
> List preparation or rental
> Brochure: writing, layout, and design
> Brochure printing and postage
> Journal advertising

Administration:
> Conference manager (correspondence, advance registration, supervision)
> Clerical personnel (mailings, registration desk during meeting, etc.)
> Conference materials (documents, badges, etc.)
> Stationery
> Postage
> Travel (conference manager, organizing committee)
> Living Expenses (manager and organizing committee; rooms and meal(s))

Speakers and Chairpersons:
> Fees
> Travel
> Living expenses (rooms and meals)

Entertainment/Recreation:
> Opening reception
> Afternoon tour
> Evening banquet

REVENUE
Registration fees (XXX @ XX) =   X

Remainder to be provided by other sources   X

If all, or a portion of, the expense is to be covered by registration fees, it is better to set your fee on an *underestimate* of attendance. An optimistic forecast of attendance that results in a low registration fee can lead to embarrassing developments, financially and otherwise. If attendance exceeds your expectations, it is easy to make the necessary adjustments, just as long as the number of registrants doesn't exceed the capacity of your meeting room!

Letters of Invitation—Recruiting Speakers

The opening approach to inviting outstanding people to a meeting is better done informally, usually by telephone. This is referred to as "feeling them out," or perhaps as an "informal inquiry" to determine whether they would be available on the date set and be willing to entertain an invitation to appear on the program. These preliminary inquiries should, of course, be made by the person on the committee or in the laboratory who is best acquainted with the one to be invited. If it is a very famous scientist, the head of the laboratory should call, except in unusual circumstances where an intimate friend or relative of the scientist is on staff.

The ploy of hinting that other well-known people will be present may work after two or three calls and no absolute turndowns, but the first call may be difficult. The best one can do is to say, referring to the tentative program, "We are also asking Carl Sagan and Jonas Salk." If the question is asked, "Have they accepted?" you can say: "We have not yet had their reply." If the response is generally positive, that is, agreement to consider an invitation and confirmation that the date is open, the next call is easier.

The person making the call should be knowledgeable about the objectives of the conference, the work of the scientist being invited, and the contribution he or she is expected to make. Although those extra incentives like climate and recreation facilities may make some difference, in most cases the final decision will be made because of the subject matter and the quality of scientific exchange promised.

After informal negotiations have indicated which speakers are likely (or certain) to accept the invitation, a formal letter must be sent to each one by the chair of the organizing committee or the head of the laboratory, setting forth the laboratory's commitments regarding reimbursements, fees, and the like, and what is expected of the guest speaker or participant.

Letters of invitation contain information about the time, place, and inclusive dates of the meeting, making it clear when the participant will be expected to arrive and to depart; whether the invitee will present a suggested topic or one "of your own choosing"; and whether a question-and-answer period will follow the presentation and, if so, if it will be conducted by the speaker or someone else. If the proceedings are to be published, this should be stated, and the due date for a written version of the presentation given. The obligations of the host institution regarding reimbursement for travel, accommodation expenses, and payment of an honorarium are spelled out unambiguously. If there is something special about the laboratory (such as its history, geography, or recent achievements) that may also be noted.

If the conference is to be very large, particularly if it is international in scope, special stationery may be necessary with the name of the meeting, dates, location, and committee members' names imprinted. This ensures that certain basic facts are not omitted. Such stationery is also useful and time-saving for the committee and the conference manager in handling general correspondence about the meeting. In most cases, however, the official stationery of the laboratory is sufficient, but letters must be carefully drafted to include all important facts, especially commitments to be undertaken on both sides. A letter that may be used as a guide can be found on page 87.

General Announcement and Invitation

As soon as the logistics for your meeting have been set—time, place, site, topic—announcements should be placed in appropriate scientific journals and other publications likely to be read by scientists interested in the subject. Even if registration will be limited because of facilities or topic, these announcements will alert other groups that those dates are taken, perhaps preventing another group from planning a conflicting meeting. Some sample announcements are shown on page 88.

Your general announcement and invitation to the conference may be in the form of a letter with a tentative program attached, but it is often preferable to create a printed brochure comprising the invitation, tentative program, and information about registration, including forms. A suggested brochure form is shown on page 91, which can be folded to fit a No. 10 envelope.

Mailings to potential registrants must include information about the program, even though the mailing is done before every participant has formally agreed to be there. If two mailings are feasible, the first can give the bare facts of the meeting—its subject, time, and location, and an indication of the topics to be presented by speakers. Such an early announcement may also include the names of those speakers who have actually or tentatively confirmed their acceptance, together with a registration form with details about the accommodations. In this announcement it should be noted that final details and a program will be mailed later. If only one mailing is to be sent, it must include a tentative program, unless—by some miracle—the final program has been set several months before the meeting.

By the time announcements (preliminary or final) are ready for posting, a mailing list must have been compiled. The laboratory, unless it is very young indeed, will have a general mailing list, and someone (perhaps the committee chair) should review that list and indicate which names, if any, are to be omitted from the invitation list. If a small, exclusive meeting is planned, or the topic relates to a field in which a limited number of researchers are currently involved, the list must be carefully combed for those primarily concerned with the topic. If the subject matter is sufficiently broad and the conference facility has accommodations for many participants, the invitation list can be broadened to include, for example, all members of a professional society or other organization with an interest in the subject, as well as younger scientists and graduate students working in the field.

SPEAKER INVITATION LETTER

August 26, 1988

Dr. John T. Smith
Department of Biology
Technology University
Cambridge, MA   02139

Dear Dr. Smith:

Each spring since 1948 the Lincoln National Laboratory has
hosted a basic research symposium on a topic of timely and
provocative interest. The overall objective of these symposia
is to summarize the most recent advances and point the way
toward further progress.

The topic selected for the 1989 meeting is Molecular and
Cellular Mechanisms of Cancer. It will be held April 6-9 at
the Stevens Hotel in Mountain View, Tennessee, a resort town
adjoining the Great Smoky Mountains National Park.

On behalf of the Organizing Committee, I am pleased to invite
you to speak at the session, "Cellular Responses to Mutagenic
Agents." A suggested title for your talk appears in the
enclosed program. Please feel free to make any changes you
would like to appear in the printed program to be sent out
with the general invitations in the fall.

Abstracts will be requested in January for use in the conference
program booklet, and manuscripts will be due at the conference.
Premium Press will publish the proceedings.

We will, of course, reimburse your round-trip travel expenses
to and from Mountain View and your living expenses while attending the meeting, in accordance with existing policy.

I hope your schedule will permit you to contribute to what promises
to be an exciting and timely symposium. If you have any specific
questions concerning the meeting or its program, please call or
write Robert F. Jones, Chairman of the Organizing Committee,
(555) 570-1200.

Sincerely yours,

Mary R. Johnson
Director

Enclosure

CONFERENCE ANNOUNCEMENTS IN PERIODICALS

The skeleton budget has an item for "List preparation or rental" for very large meetings with subjects that may be of interest to scientists who do not appear on the laboratory's general mailing list and are not likely to be known to the committee individually. Such lists may be compiled in-house or purchased from professional vendors. If the organization of the conference is being handled by outside professionals, they may be able to advise you about appropriate mailing lists that are commercially available.

Some names will already have been suggested during the program preparation and the research staff can be asked specifically for names, addresses, and research interests of others to be added. If this is your laboratory's first conference, this list may become the nucleus of a major mailing list. Every institution should build and maintain a good general mailing list, coded for selected mailings of news announcements, fellowship notices, research publications, invitations to conferences and ceremonial events, and other communications. Your database management file on your personal computer is ideal for these kinds of functions.

A general mailing to invitees should go out between three months and six months prior to the meeting. If two mailings are sent, send the first one six months before and the second as soon as the program is complete. But for a meeting of any size, final invitations must go out at least three months ahead—six months is better. Although some people hesitate to commit themselves to attend a meeting far in advance, you can offer to refund registration fees for those who may cancel by a specified date. Preregistration will immeasurably reduce the confusion at the registration desk when the conference opens, and it will also help estimate the attendance—a very important advantage if registration fees are to cover all, or part, of the cost.

Every letter or brochure must contain registration details, including fees and any extra charges for late registration, room rates, forms and instructions for making reservations, and other pertinent information about the conference site.

Registration

Scientific conferences are not rated as successes or failures on the basis of the registration process, but an intelligent, helpful, and good-natured staff at the information/registration desk contributes greatly to the smooth functioning of the meeting and to a lasting favorable impression after it is over.

Conference managers handle all preconference correspondence and advance registration. By the time the meeting opens, they should have a plan for staffing the registration desk, beginning well in advance of the first session, and continuing during coffee and meal breaks and allowing a reasonable time following each day's final session. The staff should be briefed on their duties and be able to answer questions or direct conferees to the appropriate place for answers. They must be familiar with the hotel or motel services, such as travel arrangements, medical or first-aid facilities, emergency evacuation procedures, and so on.

The registration desk performs liaison functions between conference personnel and the hotel staff and relays any important information. It should be set up near the meeting room, with a hotel telephone extension to which calls for conference registrants can be directed. A bulletin board should be placed near the desk for personal messages and general announcements.

Another bulletin board that is often the center of attraction at meetings is the job-placement board. It is wise to place this board in the general vicinity of the registration desk, but since the traffic around it may be heavy, there must be ample space surrounding it.

In addition to the extension telephone, it is a good idea to have a direct line installed for the conference manager to make calls without going through the hotel system. This is particularly necessary if the meeting is being held at a distance from the host institution, whose researchers are away from their labs all day; otherwise, many of them may be tempted to duck out to check on their experiments.

For a meeting that begins early in the morning, the registration desk should be open and staffed on the preceding evening for several hours. If most attendees have preregistered—a highly desirable situation—their conference materials can be at the desk, in personally addressed packets, arranged in alphabetical order. Each packet should contain a name badge, copy of the program, registration receipt, tickets required for special events, location of facilities in the area such as restaurants and theaters, and copies of press releases about the meeting. The issuance of packets will indicate exactly who has arrived, and those packets remaining will indicate the no-shows. The conference chair will want to keep an eye on the packets, particularly to be sure that the first speaker of the first session arrives on time.

The conference manager must be present for the bulk of the registration. After that, the front desk of the hotel or the person on duty at the registration desk must know where he or she can be reached throughout the meeting. When the first session has begun, and most or all of the registrants have arrived, one person may be able to handle the desk, but staff should be large enough to schedule each person for stints of not more than three or four hours. The desk must always be attended during breaks, since that is when most conference attendees will need information or assistance.

Publication of Proceedings

The decision must be made at the early planning sessions as to whether the proceedings are to be published and in what form. If the decision is to publish, another financial element enters the picture. Financing the publication of the proceedings of a scientific conference can be quite

CONFERENCE BROCHURE

11TH INTERNATIONAL CONFERENCE ON ATOMIC COLLISIONS IN SOLIDS

Atomic Collisions in Solids as a subject deals broadly with the fundamental aspects of particle interactions with matter in a condensed state. It crosslinks not only the basic fields of elementary particle, nuclear, atomic, molecular and solid state physics but also topical areas such as surface chemistry and physics and radiation effects on matter.

The International Conference on Atomic Collisions in Solids, held every 2 years since 1965, is one of the few conferences that truely bridges these disciplines. The Conference addresses the scientific underpinnings of such important technologies as submicron devices and new materials used in a variety of applications.

Topics that will be addressed at the 11th Conference include:

- the use of atom and ion scattering methods to probe structure, dynamics and electronic processes in solid surfaces and thin films;
- the fundamental aspects of radiation damage, sputtering, surface analysis and ion beam mixing;
- general aspects of the penetration of ions in solids (energy loss, charge states, channeling etc.);

- secondary emission of electrons from surfaces and desorption of ions and neutral species induced by electronic excitations at the surface.

For the eleventh in the series, the Organizing Committee intends to follow the finest traditions of the Conference in promoting the free exchange of ideas, providing a conducive environment for active discussions and cross fertilizations, and highlighting areas where new developments have occurred recently. The coverage will also be expanded to call into attention new and exciting work, ranging from surface scattering processes with sub-electron volt ions to elementary particle processes assisted by the crystal at hundreds of GeV.

A Call for Papers will be issued early in January 1985.

To receive that mailing, and additional information, fill out the attached card, and return it to: Kathy C. Stang, U.S. Department of Commerce, National Bureau of Standards, B348 Materials, Gaithersburg, Maryland 20899.

For further Technical Information, call Dr. P. A. Treado (202) 625-4144.

For further General Information, call Mrs. Kathy Stang (301) 921-3295.

11TH INTERNATIONAL CONFERENCE ON ATOMIC COLLISIONS IN SOLIDS

Interest Card

NAME: _____

AFFILIATION: _____

ADDRESS: _____

_____ ZIP: _____

☐ I am interested in attending the 11th International Conference on Atomic Collisions in Solids; send me more information as it becomes available.

☐ I may be interested in submitting a paper.

MAIL TO:
Mrs. Kathy C. Stang
U.S. Department of Commerce
National Bureau of Standards
B348 Materials
Gaithersburg, Maryland 20899, U.S.A.

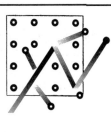

complicated and much depends on the nature of the conference, the stature of the host institution, and the relationship that exists or that can be established between the laboratory and a scientific publisher. Scientific journals devoted to a special field sometimes have an interest in printing the proceedings of a meeting, and making a cooperative arrangement with one is a relatively simple way to ensure they are done well.

If a topic has sufficient general interest, a scientific publisher may be willing to bring it out in a separate volume, but in nearly all cases a subsidy will be demanded by the publisher. The obvious nucleus of a market for the proceedings is the conference itself. Those who attend have an interest in the subject; if they participated either as speakers or in the general discussion, there is an extra incentive to buy the proceedings.

If the topic is especially exciting and the speakers are well known in the field, a publisher may be willing to forego the subsidy, and in some cases even provide an advance on royalties or a grant for clerical and other duties associated with the preparation of the manuscript. For conferences held annually, a series might be established to publish the proceedings on a regular basis.

In some cases, the cost of the published volume is estimated in advance and the registration fee includes an amount to cover the cost of a copy; each registrant automatically receives one when the publication appears. Such advance payment through registration will help convince the publisher that it might be worth doing.

When publication is planned, the first approach to potential speakers and the formal invitation sent to them should clearly state that they will be expected to submit their presentations in written form, either at the close of the meeting or by a set deadline. Even if all sessions are to be tape-recorded, speakers should submit their presentations in written form. Otherwise, some will arrive with a box of slides and a few notes on a three-by-five card, and the chance of getting a paper from them later is somewhere near zero. A tough, but highly successful, ploy is to require speakers to turn in an acceptable manuscript as a condition for receiving payment of honoraria or fees.

All sessions should be taped, since it is the only way to record discussions. Although some of the recordings may be worthless, they can become critically important when questions arise concerning who said exactly what. Moreover, speakers often extrapolate from the prepared text or insert spontaneous remarks that are as significant, or even more so, than their written presentation. But never think of trying to reconstruct accurately a whole meeting from tape recordings alone; extraneous conversation and activity in the meeting room—to say nothing of the unpredictability of technical sound equipment—make it too risky to rely solely on recordings to furnish material for publication.

No matter how elaborately you explain to each speaker that a written paper must be submitted, there will still be some who are very late in sending them, and occasionally no paper is ever sent. All you can do is set a deadline and omit any paper that has not arrived by then. The joy of publishing and the pressure to do so are strong incentives, so this will rarely occur, particularly if you set an early precedent of standing by the deadline and refusing to hold up the publication for the laggards. In a real emergency, and if a paper is especially significant, the editors can use the tapes made during the presentation to reconstruct the paper and publish a note saying it was obtained from the taped record.

Editing scientific papers can be greatly expedited if an editor (or two or three as available) attends the meeting. Editors and speakers can meet and discuss papers and establish working relationships that go far to produce a better final product. Some papers given in the first part of the meeting may even be edited during the remainder of the meeting, allowing speakers and editors to go over them before the conclusion of the conference. Even if that is not possible, the fact that they have met and talked will make later telephone communication or correspondence go more smoothly. The publication of conference proceedings is difficult and often a lengthy process, and everything possible should be done to speed it up and improve the quality of editing. You want to be proud of the final product and so does everyone whose paper or comments are included. It is probably unwise for a laboratory to attempt publication of the proceedings of its first research conference. After you have established a reputation for organizing timely conferences on subjects of contemporary and future relevance where outstanding speakers discuss significant work in the field, finding an interested publisher is easier.

The Evaluation

Following large conferences, and particularly after the first one an institution holds, there should be a follow-up evaluation to analyze and reinforce those aspects that worked well and to identify and change those that did not.

One tool that is sometimes used in evaluation is a questionnaire that participants are asked to complete at the end of the meeting. If this method is used, the questionnaires should be brief, the questions should be simple and specific, requiring only a check mark or a one-word reply, plus space for comments. The questions should concern substantive matters—the quality of the presentations, value of the meeting to the field, quality of accommodations, overall efficiency of the administrative management, length of the meeting or program schedule. Questionnaires have limited value and should never be the *only* evaluation tool used. People either lean over backwards to be polite and say everything was fine, or they give vent to their frustration over one aspect (which may be quite unimportant) and blast the whole thing. However, if there are at least a hundred people present, you will receive some thoughtful and constructive comments.

All records of major conferences should be retained, including tapes of recorded sessions, registration records, correspondence, financial records, and copies of papers submitted, even if they are not published. If the manager sends bills to the accounting department for checks to be issued, a copy of the bills should be kept in the conference files. The name and address of everyone who attended (or showed some interest in attending) should be added to your laboratory mailing list, if not already there. That list can be coded for such things as major research interests, geographic area, meetings attended, and so forth.

The conference manager should prepare an initial report covering all administrative and clerical aspects of the meeting and containing recommendations for improving procedures, eliminating unnecessary aspects, and reinforcing those that worked well and ought to be continued. Such a report is useful to future conference managers, or it could serve the same manager as a memory refresher. It might also be used to settle questions that may arise later regarding transactions with the manager of the conference site, committee members, registrants, or staff members. If the accounting department was involved in making payments, the conference manager and the responsible accountant may prepare the final financial statement jointly.

The committee chair should prepare a final report of the conference, evaluating its contribution to the field, indicating specific advances in knowledge, if any, and commenting on any new research directions that may result from the presentations. It would be useful for the chair to comment upon the procedures followed by the committee in preparing the program, and to make recommendations for future committees; to assess the site that was selected and the overall management of the meeting; to comment on the length of the meeting, length and timing of the sessions, and the merits or drawbacks of the planned entertainment. It will not be easy to get the chair to do this; so the initial appointment should include a statement that such a report is required, and a date should be set for its submission as soon as possible after the meeting. The further away the meeting date gets, the more difficult reporting on it becomes. The final report should be read before a meeting of the research staff, that includes a discussion, and retained in the files along with a memo summarizing the discussion for reference by future conference planners.

Well-planned and well-managed scientific conferences of high quality are a credit to the institution and enhance its stature.

EXTERNAL COMMUNICATION: THE PUBLIC

The head of a research institution has a responsibility to keep the public informed about what is going on in the laboratory. News releases should be issued regularly about any event of importance or of interest to the public and distributed through all the media—newspapers, magazines, radio, television. On occasion, employees of your institution should be permitted to invite their families and friends to an "open house" where they may see, first hand, the mysterious inside of a science laboratory. Removing some of the mystery can go a long way toward reassuring a community that may be apprehensive about what is actually going on in "that place." Journalists and science writers must be welcome and their visits handled in such a way that they get the whole story. If they are seeking information on research about which there are conflicting views within the laboratory, representatives of all schools of thought should be interviewed.

Philip C. Ritterbush, formerly director of academic programs at the Smithsonian Institution, cautioned:

> Reason cannot be defended by sealing it off in sanctuaries of intellect. Unless the sanctuaries are thrown open, they may be invaded and the darkest of prophecies may be fulfilled. . . . Reason's best champions will be those who frame new public contexts for the dissemination of knowledge, where citizen interest can be engendered, where better informed views can be nurtured, and where any voice that is raised can be assured a hearing within the technical community and a response from some appropriate educational or governmental institution.[3]

Another view on the same point was expressed by Carl Sagan in an interview with Boyce Rensberger of The New York Times:

> There are at least two reasons why scientists have an obligation to explain what science is all about. One is self-interest. Much of the funding for science comes from the public and the public has a right to know how their money is being spent. . . . The other is that it's tremendously exciting to communicate your own excitement to others. . . .[4]

Although public perception of a research institution will be formed mainly by its scientific contributions, there is much that can be done, and needs to be done, to enhance public esteem for each research institution, and ultimately for the whole scientific community. The fact that most research is supported by public funds is only one of the reasons. Another, more urgent, reason is to encourage young people, very early in their academic lives, to be aware of the world of science and the many opportunities it offers them to develop their talents and to contribute to advances in knowledge.

A certain amount of interaction with the community in which a laboratory is located will take place haphazardly. A researcher may teach a course at a local university or junior college, become active in the Scouts, or affiliate with a camping, hiking, or bird-watching group. Persons who encounter staff members through these activities will form an impression about the institution, but their number is so limited that it is not enough. A serious effort has to be made to acquaint the community with the lab's significance and with the doors of opportunity it may open.

When there is never enough money for research, it may seem like a frill to suggest that the budget include funds for anything so seemingly irrelevant as "public relations," but unless responsibility is fixed for handling matters that fall into that category a great deal of research staff time will inevitably be siphoned off into dealing with them.

When the budget is too small to allow for a full-time staff (or even one person) to handle public or community relations, responsibilities can be distributed among administrative staff members, or may be handled by committees. Then, when a question or a need arises, there is a designated person or office to which it can be referred. Every laboratory will have its own individual requirements and strategies for dealing with the public.

Brochures and Pamphlets

Brochures and pamphlets describing the work going on in the laboratory should be written in language the layman can understand. They should give the facts, in readable prose, that every visitor, prospective employee, researcher, member of the administrative staff, consultant, or program reviewer will find informative or useful. It is a good idea to include a short history of the lab, its general purpose, and a breakdown of the research staff by individual laboratories, as well as the names of the senior researchers and the support staff. If the laboratory is spread out over a large area, a map indicating the location of each laboratory is also helpful, particularly for visitors.

When he was associate director for genetics at the National Institute of Environmental Health Sciences at Research Triangle Park, North Carolina, Frederick J. de Serres, said:

> Give every visitor something to take away that describes the work going on in the laboratory. It does not have to be a slick, four-color professional brochure, but it should be well written and describe the kinds of research activities the lab is engaged in.

The laboratory's publications list may be a part of the descriptive brochure in the beginning, but after a few years, the productive laboratory will have a list that is extensive enough for a separate document. If there is an editor who assists with manuscripts and maintains records on publications, the running record will contain all the necessary information, and the list will need only to be updated when necessary. No other document will give so clear a picture of the overall research activity of a laboratory as its publications record for recent years.

These brochures should be designed to fit into a No. 10 mailing envelope, so they can be included with letters to outsiders who have an interest in the work of the laboratory, and should be enclosed as accompanying documents with grant applications. They must be updated regularly.

News Releases

News releases let the public know about important research findings, of course, and there are also many other events that warrant public announcement: appointments of key staff members; travel by the research staff to present papers at other institutions or major research conferences; honors received by staff members, or their election to important scientific society offices; appointments to governmental study panels or national committees; announcements of important patents, grants, or contracts received. Large research conferences or annual society meetings held at the institution generate a lot of public interest, particularly when they bring well-known scientists into the area, and especially if any of the lectures or other program events are open to the public. An example of a well-written news release appears on page 98.

Tours and Visits

The laboratory workplace is seldom of great interest to the average sightseer, and most scientists do not take kindly to the idea of showing visitors around. However, college, university, and even high school students who experience firsthand the atmosphere of a scientific laboratory and talk with people directly engaged in research, gain a perspective not otherwise possible in their education. They may be exposed to an aspect of science that arouses their enthusiasm for research in a particular field or inspires them to press ahead with an already established goal. Every laboratory, in the interest of encouraging future generations of scientists, has a responsibility to welcome such visits.

Government laboratories, or any institution that applies for a government grant, can expect to be site-visited from time to time, or to be faced with a congressional or other type of delegation. And there are always visits by review committees, and candidates for staff positions. While the success of these visits is the responsibility of the research staff, they may need assistance from experienced public-relations personnel.

"Open house" tours should be held from time to time, perhaps annually, when families and friends of the entire staff are invited to see where their relatives and neighbors work.

All visits to the laboratory must be conducted in a manner that minimizes interruption of on-going research while it maximizes the benefits to the visitors. In order to accomplish this, someone must be in charge of planning, organizing, and conducting tours, including establishing safeguards against every conceivable, and some nearly inconceivable, event such as injury, theft, and property damage.

No visitor should go away empty-handed. The printed brochure describing the work of the lab and showing the organization of the research program can be given to everyone, along with other appropriate

NEWS RELEASE

```
FOR IMMEDIATE RELEASE
---------------------

Lafayette, N.Y. - Final preparations are being completed for
the IV International Congress of Environmental Science, which
will meet August 10-16 in the United States for the first
time in its 20-year history.
        The meeting will bring to the campuses of the State
University of New City College of Ecological Science close
to 2,000 of the world's preeminent ecologists including
biologists, botanists, foresters, chemists, statisticians,
landscape architects and others whose work focuses on the
environment.
        A number of internationally prominent scientists will
speak on topics ranging from acid deposition, desertification,
the shrinking of the world's tropical forests, and toxic
waste management, to the increase of carbon dioxide in the
atmosphere, soil fertility, long- and short-term evolutionary
constraints, and industrial stress on ecosystems.
        Included in the program are Eugene F. Smith of the
University of Washington, David Jones of Cornell University,
Harold Johnson of Stanford University, Robert Rogers of the
University of Washington, Gene March of the Carey Arboretum,
and George Walsh of the Environmental Defense Fund.
        Roughly a dozen concurrent program sessions are planned
for each morning and afternoon the conference is in session
and will be held in a variety of locations on the campuses.
        The International Congress of Environmental Science
is the quadrennial meeting of the International Association
of Ecoscience, an umbrella organization of nongovernmental
scientific associations as well as international thematic
groups in the field of ecology.
        Dr. Robert L. Bond, chairman of the Faculty of Environmental
Biology, and Dr. Paul Henry, professor of Environmental
Biology, are the Congress chairmen.
```

documents, such as current news releases of important events, or reprints of recent publications.

Special Events

Ceremonial occasions include the awarding of prizes or honors to the institution's staff, to others by the institution, or the celebration of a milestone, such as the lab's twentieth or one hundredth anniversary. Seminars, conferences, or lectures organized by the scientific staff are often concomitants of such events, but someone has to make arrangements for the formal ceremonies and attendant social entertainment. Even if a special committee is set up to handle each separate event, all plans must be channeled through a central office to make certain that nothing has been overlooked or duplicated.

INTERNAL COMMUNICATION

It is an embarrassing, but not a rare, occurrence for a scientist away on a visit to another lab to hear about an honor bestowed upon his colleague back home or about work going on in his own institution that he or she had not previously known about. It is important that members of the staff be kept fully informed about matters of concern to them. This includes the free exchange of scientific information among the research staff as well as the circulation of pertinent information of a general nature to everybody routinely, or if the occasion warrants it, hastily.

In small, recently organized laboratories where staff members see each other every day, a certain amount of information is exchanged during chance meetings in hallways, at the cafeteria, or along the always-functioning grapevine, but these channels are not dependable, and they are soon overloaded and outgrown. A well-planned, properly administered system of internal communication contributes greatly to the morale and smooth functioning of the laboratory.

Availability of the Laboratory Director

Encouraging communication between scientists who may view each other as rivals is a challenge to scientific administrators, especially since some researchers may have that slight touch of paranoia that makes them suspicious of anyone, including the head of the institution, who expresses an interest in the details of their work. Recognizing the hazards, many American administrators are reluctant to follow the European custom of visiting every laboratory in the institution at least once a day. Harvey Patt, the late director of the Radiobiology Laboratory at the University of California, San Francisco, School of Medicine, used to say that he counted on "collision frequency" to provide him opportunities to talk with the research staff. Meetings in the elevators or hallways were, he felt, sufficient for him to keep in touch. James Clark Maxwell, the founding director of the Cavendish Laboratory from 1871 until his death in 1879, took a different view. While he continued his own very active research on electricity and magnetism during his tenure as director, he also managed to visit every research lab in the Cavendish once a day. Of course, the Cavendish was much smaller in his time than it is today. In the early days of the biology division at Oak Ridge National Laboratory, Alexander Hollaender made the rounds of every lab to say "Good morning." But as the laboratory grew to a staff of more than 600, he limited his morning stops to a few each day. Hollaender, an early riser, was always among the first to arrive in the morning; consequently, this traditional, but informal ritual came under suspicion. Some suggested he was "taking the roll," that he merely wanted to see what time they started to work.

Visits to individual laboratories, however, are more a matter of style than a means of information exchange. It provides one opportunity for the research staff to speak with the head of the institution if they wish to do so, but that can be accomplished in other ways, such as maintaining an open-door policy.

At the Burroughs-Wellcome Research Laboratories, Robert A. Maxwell, head of the pharmacology department, was asked how often he sees the president of the laboratories. He replied, "Not very often. Nor do I need to see him very often. But he is available and I know that I can see him if I need or want to." Even if the head of the institution seldom sees and talks with each staff member, information on the research activity and progress of each is routinely channeled (or should be) through publications, progress reports, and seminars. The administrator of a research institution must resign himself to spending a great deal of time reading and attending seminars in order to be fully aware of the nature, quality, and status of the institution's research activity.

Some successful laboratory administrators have created social occasions where they meet the staff informally and encourage an easier conversational exchange than is possible surrounded by the normal trappings of the laboratory.

Professor (Lord) Ernest Rutherford, under whose leadership from 1919 until 1937 the Cavendish Laboratory reached its apotheosis, and his wife, Lady (Mary Newton) Rutherford, held regular Sunday afternoon tea parties at their home for the staff. And at biweekly meetings of the Cavendish Physical Society, Lady Rutherford presided over the teapot that was set up between the lecture bench and the blackboard. Lord Rutherford took a very personal interest in his researchers and felt pride in their accomplishments, and his concern for their welfare did not stop at the laboratory gates. During one of the regular Sunday tea parties, Rutherford took aside the recent bride of a young, brilliant Cavendish physicist and explained to her that physicist husbands must be freed from all household chores because they had to devote themselves entirely to their work.[5] One wonders what he would say today to the husband of a brilliant young woman physicist.

Alexander Hollaender was frequently described as "Herr Professor" and his German accent, which he never lost, reinforced the image. His strong personality, together with his reputation as a demanding administrator, caused young researchers to approach him only when absolutely necessary and then with a certain amount of dread; that is, until they got to know him. The Hollaenders also entertained the staff at their home, not on a regular weekly schedule, but upon any occasion that might conceivably serve as an excuse for a party, and it could happen several times in one week. Engagements, weddings, or the birth of a baby were as likely to provide such an occasion as the visit of an eminent scientist from abroad—and there was an abundance of all such events. Everybody on the staff, including student trainees, postdoctoral fellows, and laboratory assistants, was invited at one time or another and, in turn, they invited the Hollaenders. They went hiking with him, named their babies after him, and even got married in the Hollaender's house.

Every Sunday morning, for more than twenty years, in winter and summer and through rain, sleet, and snow, Hollaender led a group into Cumberland Mountain terrain devastated by strip miners, where they picked up prehistoric fossils that the earthmovers had brought to the surface. Samples of the Tennessee fossils today can be found all over the world in the homes and offices of scientists who went along on those hikes. Conversations on those hikes ranged widely, but scientific matters predominated, and a great deal of information exchange and collaborative work resulted.

During World War II gas-rationing, the only public transportation available to and from one of the wartime labs was a daily bus that went to the installation once in the morning and returned to the residential area again in the evening. Katherine Way, a physicist who rode those buses, says, "It was one of the most fruitful sources of exchange of ideas, that half-hour or forty-five minutes twice a day which scientists were forced to spend together." Such opportunities for informal exchange may be more likely to occur in somewhat isolated places like Oak Ridge, where productive unplanned contacts can arise out of circumstances of time and place. But they can also occur in large European cities. The Polish mathematician S. M. Ulam wrote:

> These long sessions in the (Lwow) cafes with Banach and Mazur were probably unique. Collaboration was on a scale and with an intensity I have never seen surpassed, equaled or approximated anywhere—except perhaps at Los Alamos during the war years. . . . There would be brief spurts of conversation, a few lines would be written on the table, occasional laughter would come from some of the participants, followed by long periods of silence during which we just drank coffee and stared vacantly at each other.[6]

Seminars

Intellectual cross-fertilization and information exchange cannot be left entirely to happenstance in a research institution, but must be encouraged through a planned program of internal seminars, formal and informal.

During the early days of the biology division at Oak Ridge, there was no cafeteria or nearby restaurant, so everybody brought a lunch from home. Brown-bag seminars arose spontaneously as small groups formed in various labs during the noon hour where they ate lunch together and discussed their work. The idea gained popularity and eventually the informal seminars became systematized. A small conference room was set aside and each day of the week was designated for discussion of one area of research—genetics on Monday, biochemistry on Tuesday, and so on. Scientists working in each designated area were responsible to provide a speaker or discussion leader for their day of the week. There was never any attempt to make attendance at these sessions compulsory, but everyone in the laboratory was welcome, including nonscientific staff.

Scientists found that aside from providing a common lunchroom, these seminars eliminated the midday break in concentration, and the momentum of the morning thought and activity was continued, or sometimes was accelerated, by the discussion. If the morning had been disappointing, the session might provide stimulation for the afternoon.

European visitors were privately shocked by this irreverence for the ritual of the midday meal, although they were too polite to say so openly. Some confided to close associates that they felt the break of an hour or preferably two hours when one could go out to a restaurant for a hot lunch was a most civilized custom. But during the past few years, rumors have been spreading that the brown-bag seminar has become popular in some French (French!) laboratories, and may be gaining in other countries also.

Young scientists who are just beginning to be invited to give papers at conferences and society meetings can gain confidence and improve their presentation techniques at these informal seminars. Pre- and postdoctoral students can receive valuable platform experience and training in presenting their work to an audience. These informal gatherings are an excellent source of information for the laboratory head who wants to keep up with the research program and are also useful as one means of evaluating prospective staff members. A curriculum vitae may list published work, but usually does not indicate a scientist's current projects and plans for continuing or extending that work in the future. Participating in a seminar also reveals one aspect of the personality of the prospective colleagues as an indication of their ability to fit into the group personally as well as professionally.

Seminars by outside visitors may be somewhat more formal, and it is better to schedule them at the end of the day, perhaps an hour before the workday normally ends. People can then plan to come to a stopping point an hour earlier, rather than breaking up their working day (although many will return to the laboratory after the seminar). For a sample announcement of a seminar, see pages 103–104.

These slightly more formal seminars are also one way to recognize staff members whose work has recently yielded outstanding results by asking them to discuss it before the entire staff. It is also a good way for researchers to keep up to date with the productivity and range of activity throughout the institution.

If something special is accomplished in an area of interest to many scientists in various disciplines, an occasional evening lecture may be

INTERNAL BULLETIN

NIH CALENDAR OF EVENTS

The Calendar of Events, issued each Wednesday, lists NIH-sponsored meetings. Other meetings sponsored by local scientific and educational organizations may be listed if devoted to subjects related to NIH missions and if space is available in the Calendar. Please send notices for the Calendar to Betty MacVicar, Editor, Bg. 31, Rm. 2B03. Notices must be in writing and must be *received no later than 10 a.m. on Tuesday of the week preceding the meeting.* For further information, call 496-2266.

April 6–12, 1987

NIAID KINYOUN LECTURE. AN NIH CENTENNIAL LECTURE.
REGULATION OF IMMUNITY BY CLASS II MHC ANTIGENS.
Speaker: Dr. Hugh O. McDevitt, Professor & Chairman,
Dept. of Med. Microbiology, Stanford Univ. Sch. of Med.
Fri., April 10, 3 p.m.— Wilson Hall, Bg. 1. (Info.: 496-5717)

Mon., Apr. 6

8:30 a.m.–5 p.m.   NHLBI/NICHD/HRSA/OMAR Consensus Development Conference. NEWBORN SCREENING FOR SICKLE CELL DISEASE AND OTHER HEMOGLOBINOPATHIES. *Also on Tues., 8:30 a.m.–12 m. and Wed., 8:30–10 a.m.* Bg. 10, Masur Auditorium. (Info.: 496-4236)

10 a.m.   NINCDS-LMB Seminar. INTRACORTICAL MICROCHEMISTRY IN THE BRAINS OF ALZHEIMER'S DISEASE PATIENTS. Speaker: Dr. Sondro Sorbi, Inst. of Mental & Nervous Dis., Univ. of Florence, Italy. Bg. 36, Rm. 1B13. (Info.: 496-5468)

11 a.m.   NCI Immunology Seminar. BIOCHEMICAL AND FUNCTIONAL STUDIES OF THE CD2 (T11) and the CD1 MOLECULES. Speakers: Alain Bernard and Laurence Boumsell, Inst. Gustave-Roussy and Hopital St. Louis, France. Bg. 10, Rm. 4B36. (Info.: 496-5461)

11 a.m.–12:30 p.m.   NCI-FCRF-LCC Seminar. CELL MEMBRANE LIPID FLUIDITY, RECEPTOR EXPRESSION, CELL FUNCTION: EXPERIMENTAL AND CLINICAL STUDIES. Speaker: Dr. Gerhard R.F. Krueger, Immunopathol. Section, Pathologisches Institut Der Univ. Zu Koln. FCRF Bg. 549, Auditorium, Frederick. (Info.: 301-698-5864 or FTS 8-978-5864)

12 m.   NIDDK-LCP Seminar. KINETICS OF DOMAIN FORMATION IN SICKLE HEMOGLOBIN. Speaker: Dr. Frank Ferrone, Dept. of Physics, Drexel Univ., Phiadelphia. Bg. 2, Rm. 102. (Info.: 496-1024)

12 m.   NHLBI-MDB Seminar. REGULATION OF ECTODERM SPECIFIC GENE EXPRESSION IN FROG EMBRYOS. Speaker: Dr. Thomas Sargent, LMG, NICHD. Bg. 10, ACRF, Rm. 7C101. (Info: 496-5095)

12–1 p.m.   NIDDK-DB Seminar. ONTOGENY AND FUNCTION OF INSULIN BINDING TO THE EARLY MOUSE EMBRYO. Speaker: Dr. Susan Heyner, Dept. of Obstet. & Gynecol., Albert Einstein Med. Ctr., Philadelphia. Bg. 10, Bunim Rm., 9S237.

12–1 p.m.   NHLBI-LBG Seminar. MOLECULAR BIOLOGY OF CATECHOLAMINES. Speaker: Dr. Jacques Mallett, Lab. of Cellular & Molecular Neurobiology, Dept. of Molecular Genetics, Centre Natl. de la Recherche Scientifique, Gif-sur-Yvette, France. Bg. 36, Rm. 1B-13. (Info.: 496-4811)

12:15–1:30 p.m.	NCI-FCRF-MMCL Seminar. DNA A FUNCTION IS REQUIRED FOR IS50 TRANSPOSITION. Speaker: Dr. Suhas Phadnis, Dept. of Microbiol., Washington Univ. Sch. of Med., St. Louis. FCRF Bg. 539, Conf. Rm., Frederick. (Info.: 301-698-5864 or FTS 8-978-5864)
1–2 p.m.	NCI-FCRF-LEI-BRMP Seminar. THE ROLE OF DIETARY FATS IN CANCER. Speaker: Dr. B. Lokesh, Cornell Univ. FCRF Bg. 560, Library, Frederick. (Info.: 301-698-5864 or FTS 8-978-5864)
1:30 p.m.	NINCDS-LMG Seminar. SOME OBSERVATIONS ON PROTEOLIPID PROTEIN (PLP) AND MYELIN IN THE MYELIN-DEFICIENT RAT AND JIMPY MOUSE. Speaker: Dr. Ian Duncan, Sch. of Vet. Med., Univ. of Wisconsin. Bg. 36, Rm. 1B07. (Info.: 496-9106)

Tues., Apr. 7

9 a.m.	NIAID-AIDSP Seminar. A NEW FELINE T-LYMPHOTROPHIC RETRO-VIRUS. Speaker: Dr. Niels Pedersen, Univ. of California at Davis. Bg. 31, Conf. Rm. 9, C Wing, 6th Floor. (Info.: 496-0545)
9:45 a.m.	NICHD EEO Annual Meeting. FILM: "THE CONSTITUTION: THAT DELICATE BALANCE—AFFIRMATIVE ACTION VS. REVERSE DISCRIMINATION. Bg. 1, Wilson Hall, 3rd Floor. (Info.: 496-5133)
10:30–11:30 a.m.	NICHD-LTPB Seminar. EFFICIENT AND ROBUST DESIGN OF EXPERIMENTS FOR ESTIMATION OF NONLINEAR PARAMETERS. Speaker: Dr. Lazlo Endrenyi, Dept. Pharmacol., Univ. of Toronto. Bg. 10, Rm. 4B36. (Info.: 496-6571)
11 a.m.	NCI Lab. of Chemoprevention Seminar. REGULATION OF FIBRONEC-TIN BIOSYNTHESIS. Speaker: Dr. Suzanne Bourgeois, Salk Inst., La Jolla, CA. Bg. 41, Conf. Rm. (Info.: 496-5391)
12 m.	NICHD-LNN Seminar. USES OF VACCINIA T7 RNA POLYMERASE EXPRESSION VECTOR SYSTEM. Speaker: Dr. T. Feurst, LBV, NIAID. Bg. 36, Rm. 1B13. (Info.: 496-3239)
12 m.	NIH-wide Immunology Seminar. IN VIVO AND IN VITRO FOOTPRINTING OF THE HUMAN IMMUNOGLOBULIN KAPPA ENHANCER. Speaker: Jeff Gimble, LIG, NIAID. Bg. 10, ACRF Amphitheater, 1C114.
3:30 p.m.	NIMH-OD Seminar. CONTRASTS IN THE PSYCHOLOGY OF GENIUS: CHURCHILL AND NEWTON. Speaker: Prof. Anthony Storr, Oxford, England. Bg. 10, ACRF Amphitheater, 1C114. (Info.: 496-4183)

arranged to which not only colleagues but family and friends are invited also.

It is always a good idea to follow end-of-day or evening seminars with a refreshment hour during which everyone can meet the speaker and continue the discussion.

Staff Meetings

Scientists tend to resist formal, regularly scheduled administrative meetings, and if such gatherings are poorly run and become boring, staff members can be very creative in coming up with reasons why they cannot attend: "In the middle of a very critical experiment. Sorry." There must, however, be a procedure that ensures regular contact between the staff or faculty, at least the senior researchers, and the chief administrator. The frequency of such occasions depends on the size of the institution, the number of individual research units, and the need to convey information. Meetings are useful for making announcements of interest to everyone, and for giving the senior staff the opportunity to bring up matters of individual concern.

There should always be an agenda, and anyone who attends should be invited to place items on the agenda for discussion. A deadline for submitting agenda items must be set sufficiently in advance of the meeting date for the staff members concerned to be alerted and prepared to discuss the item. The heads of every administrative unit—personnel, accounting, purchasing, publications, library, and so on—should attend staff meetings since their departmental functions may be discussed. Open expression of opinion concerning the administrative services is a good way to ensure that the needs of the scientific staff are being met. Although every laboratory will have its share of carpers who are never satisfied with the services they receive, they are soon recognized. But it is a great mistake to dismiss someone as a chronic faultfinder until thorough investigation of complaints backs up that opinion. If the administrative staff assumes a defensive position and rebuts complaints instead of investigating them, the whole exercise is useless.

Appropriate agenda items for staff meetings are those matters of concern to everyone in the laboratory, such as safety, personnel policies, distribution of supplies, general-use equipment, community relations, library holdings and service, and anything affecting the general welfare and morale.

One method of ensuring that matters of interest to everyone are placed on the agenda and at the same time increase interest and participation in staff meetings is to have a rotating chair. Chairpersons for each meeting are appointed from a roster of all staff members or, in large laboratories, it may be necessary to restrict the list to senior staff. The chair is responsible for preparation of the agenda and presides over the meeting. It is advisable for the head of the laboratory to have a standing place on the agenda for administrative matters that might otherwise be omitted.

If the meeting room is large enough, it is a good idea to invite everybody, although a skeleton staff will have to remain in some labs or departments. When the staff is too large for a general meeting, the head of each research unit or service department should attend and be responsible to convey the details of the meeting to his or her colleagues.

The conduct of a staff meeting requires not only good planning but a firm will to keep the proceedings on track; unless the chair maintains full control of the session, it can be sidetracked by irrelevant discussion. Certain scientists can be particularly difficult in insisting upon introducing subjects that are not on the agenda, and that cannot be handled without prior preparation. They are often self-appointed vigilantes who feel it is their duty to alert the scientific community to some approaching danger with the announcement that "Even as we sit here pondering these insignificant questions, our academic and intellectual freedom is being threatened by. . . ." The threat may be the platform of a political candidate, a recently proposed piece of legislation in Congress, or an editorial in the morning paper. The purpose of the gathering can be thwarted by such interruption and the meeting turned into a lengthy discussion on solutions to these threats. Decisions on the points for which the meeting was called are not reached, or are arrived at hastily. Not uncommonly, an additional meeting must be called either to reassess hasty decisions, or to arrive at any decision at all.

The vigilance and energy of those individuals must be channeled into constructive use. Immediately appoint the interrupter to head a committee to look into the matter and report to the group at the next meeting; then move on to the next agenda item. (This is an example of a situation in which the mere appointment of a committee serves a purpose.) This action will prevent the meeting from being turned into a rally for a cause instead of dealing with the items on the agenda.

It is essential that notes be taken at all meetings, not necessarily minutes that record every detail, but a record of agreements made, or commitments to investigate complaints or suggestions. And every item noted must be followed up. The credibility of the administrator and the validity of the meetings are quickly destroyed if agreements made are not fulfilled and reported on at the time set for a report.

Newsletters

A laboratory of a reasonable size, say more than fifty people, needs a regular internal information sheet, bulletin, or newsletter to keep the staff informed about recent or planned activities. Items of general interest include expected visitors, seminars, recent library acquisitions or capabilities; recent publications of significance by staff members; honors received by anyone in the laboratory, such as prizes or honorary degrees, significant travel to other laboratories or major conferences; local appearances by staff members at community gatherings or on radio or television shows; and the names of new or departing staff members. This bulletin should be reproduced very simply, preferably photocopied, to ensure that it can come out regularly on the day scheduled. It will become an aid to the staff in planning ahead to attend certain seminars, arrange to meet with visitors for scientific discussion, and keep informed about administrative facts they should know.

It should never be allowed to become a "gossip sheet," with stories about vacations, weddings, and births, but if there is a great deal of interest in some recreational activity, such as plans for an all-lab picnic or an important softball game, a brief announcement might be included.

The purpose of the newsletter is to provide information that is useful in the achievement of the laboratory's scientific goals. In some cases, it should be mailed to consultants or advisers of the lab who wish to be informed about internal events.

Bulletin Boards

Few communication tools are more effective than a meticulously maintained, current, and orderly bulletin board. But who has ever seen one? Keeping a board up to date requires daily screening and removal of passé items. Unless the board is very large and can be divided into two sections, there should be two boards: one for permanent postings, and one for current items only.

The bulletin board cannot be abolished; there are too many required postings, such as emergency telephone numbers, evacuation procedures, and notices about employee and visitor rights or protection.

If a second board is used to post current items, it must be under the control of one person who alone has responsibility to decide what items are posted, and when items are removed. In order to enforce this procedure, a glass-enclosed board that can be locked is a necessity.

In some laboratories, small boards where humorous material may be posted by anyone at will provide lighthearted moments during a monotonous task, a grueling experiment, or just a bad day. These boards are generally confined to individual labs, but they are also appropriate for the cafeteria or lunchroom and recreation areas.

Paging Systems

Even in the best-managed laboratories there are times when information must be transmitted instantly to everyone, and the only effective way to do this is with a speaker system, clearly audible in every part of the laboratory.

There also must be a system for calling staff members who may be needed urgently and immediately, such as those concerned with safety, maintenance of facilities, or medical care.

We are fortunate to be living in a time when electronic technology has made instant communication easily available. Nonetheless, overuse of speaker or paging systems is annoying and disruptive in a research laboratory, and will eventually defeat its purpose by causing people to "turn off" psychologically. The system should be used only in urgent situations.

Another caveat to keep in mind concerning such systems is that they should be only as elaborate as the institution requires. Not every lab needs the most expensive and extensive system; the design should be carefully tailored to the needs of the laboratory, allowing for anticipated growth. The indispensable element is reliability; a system that cannot be depended upon is worse than none at all.

REFERENCES

1. Robinson D.O. Where the Elite Meet. *Sciences* 18(10):18–19, 1978.
2. Davis K.A. How Your Agency Can Organize a Conference. *The Grantsmanship News* 3/4:23–34, 1985.
3. Ritterbush P.C. The Public Side of Science. *Change* 9:32, 1977.
4. Rensberger, B. Carl Sagan: Obliged to Explain. *New York Times Book Review* May 27, 1977. p. 8, 19.
5. Crowther J.G. *The Cavendish Laboratory 1874–1974*. New York: Science History Publications, 1974. p. 223.
6. Ulam S. M. *Adventures of a Mathematician*. New York: Charles Scribner's Sons, 1976. p. 34.

Chapter 7

PUBLICATION OF RESEARCH RESULTS

And differing judgments serve but to declare
That truth lies somewhere, if we knew but where.
　　　　　　　　　—William Cowper, *Hope*

Many gifted scientists find the task of writing up their work quite dismaying and there is a great deal of poorly written scientific literature to prove it. Moreover, inspired researchers often resent taking the time to write about something that is finished when they could be getting on with their next idea. In his excellent book, *How to Write and Publish a Scientific Paper*, Robert A. Day introduces his preface with the following quotation from Sir James Barrie: "The man of Science appears to be the only man who has something to say just now—and the only man who does not know how to say it."[1] It is certain from the body of Sir James's work that he was using the word "man" to stand for "human," and it is also true that scientists' ability to "say it" has improved somewhat since that was written early in the twentieth century. But since every scientific discovery builds upon the past work of others and the future of science rests upon the continuation of this process, no piece of scientific research has been completed until it has been communicated.

ENDORSEMENT OF MANUSCRIPTS

The laboratory has a responsibility to see that the results of research are disseminated, and in this connection it has the greater responsibility to see that the work is of a high quality and accurately reported. This is perhaps the weightiest and, at the same time, the most delicate matter a scientific leader has to deal with in the management of a laboratory. One school of thought holds that scientists are responsible for their own work and each must stand or fall alone; that the institution has no responsibility or right to decide whether a scientific communication should be issued. Others are a bit more conciliatory and allow for a certain degree of institutional oversight, but view the right to disapprove the submission of a paper to a journal or a book to a publisher as censorship.

The fact remains that the research institution whose name is associated with a publication has a responsibility to the public, to current and future scientists, and to its own institutional future to ensure that publications emanating from the laboratory meet the following criteria: the work is of a high quality, scientifically accurate, and truthfully reported, and that credit is given fairly to those who performed it and to those whose work it was built upon. Only by establishing and maintaining these standards can research directors fulfill their responsibility to see that work produced under their leadership conforms to the highest standards of scientific inquiry.

When Arthur Miles, the scientist in C. P. Snow's novel *The Search*, failed to gain an important post because of an error his assistant made which resulted in a false scientific report, Snow has Miles's mentor, Professor Hulme, say to him:

> Now if false statements are to be allowed, if they are not to be discouraged by every means we have, science will lose its one virtue, truth. The only ethical principle which has made science possible is that the truth shall be told all the time. If we do not penalise false statements in error, we open up the way, don't you see, for false statements by intention. And of course a false statement of fact, made deliberately, is the most serious crime a scientist can commit. There are such, we both know, but they're few. As competition gets keener, possibly they will become more common. Unless that is stopped, science will lose a great deal. And so it seems to me that false statements, whatever the circumstances, must be punished as severely as is possible.[2]

In recent years, that "most serious crime" has, indeed, become more common; several scientific institutions and universities in the United States have besmirched their honor and, to a degree, the honor of the scientific profession by putting out false scientific data. In some cases, it was a deliberate attempt to influence the distribution of research funds in order to continue the work, perhaps with every intention of achieving the results that were reported as already having been obtained.

It is not easy to prevent false reporting, whether it occurs intentionally or through honest error, but the most effective insurance against dissemination of inaccurate information is a rigorous review procedure. Every report of scientific progress or achievement that is officially submitted by a staff member should be reviewed by the laboratory leader or someone else with the ability to attest to its quality and accuracy. This includes not only those papers submitted for publication, but also progress reports that accompany budget requests or grant applications and requests for grant renewals.

The institution head rarely has the detailed knowledge to vouch for all the work, and two or three qualified staff members or outside experts must be given responsibility for reviewing work in particular areas.

The charge to reviewers should include these criteria: Is the scientific level of the work appropriate for the institution? Was the work performed in accordance with the best scientific methods? Is it being reported accurately and are the conclusions or interpretations logical and well stated? Have appropriate citations of the work of others been included? Has credit been given to all who contributed substantially to the performance of the work? Finally, is it readable and understandable? Reviewers should be encouraged to discuss with the author any point that seems in doubt. It is preferable to have more than one reviewer for each manuscript; this provides a wider range of opinion and there is less likelihood that the review will carry a taint of personal bias.

If the majority of the reviewers feel that a report or paper does not meet the established standards for submission, and the author refuses to make the suggested changes or to withdraw the paper, the head of the laboratory must handle the matter. In highly controversial cases, it might be advisable to ask for a further review by a disinterested outsider, or more than one, with knowledge in the field. If the outside and internal reviews agree, the head of the laboratory has no choice but to refuse the laboratory's acknowledgment and endorsement of the paper.

Outside reviewers may also be called upon if there is no one in the institution qualified to make a determination about the work, as sometimes happens when a researcher is moving into a new field, or when there is a limited number of scientists at the laboratory working in a particular field. Some laboratory heads and some individual researchers have reciprocal arrangements with colleagues in other institutions whereby they provide each other with reviewing services as needed on a complimentary basis. This service can also be obtained through the appointment of outside consultants with expertise in particular fields of science.

In the twenty years Alexander Hollaender headed the biology division of the Oak Ridge National Laboratory, he refused to allow a scientific paper to be submitted less than half a dozen times; he says today that he regrets none of those actions. In every case, he believes subsequent events proved the decision to have been right.

The size of the laboratory and the number of researchers in the various fields will determine the number of reviewers needed and their responsibility. In a modest-sized place with few researchers, reviewers might be able to offer substantial help in revision. Reviewers will often be senior scientists with respectable publication records and their advice can be valuable to less-experienced researchers. The dangers to be avoided in any review system lie in the extremes. No system should become a censorship device. But the best review system in the world is worthless if it becomes a rubber-stamp operation.

There are two highly important considerations that must be taken into account in naming internal reviewers or arranging outside review: one is confidentiality and the other is rivalry. Reluctance to write up experiments stems from several sources, but a major source is the researcher's fear that the work will be jeopardized by revealing the details to others at an early stage. Sometimes it stems from insecurity about the quality of the work or the accuracy of the interpretations, but more often from fear of being "scooped" by others working on the same problem, or of being attacked by rivals who favor another approach or theory. It is commonly supposed that this resistance is more likely to be encountered in industrial labs than in others; therefore, the comments of James B. Fisk when he was president of Bell Telephone Laboratories are significant:

> Freedom of publication and encouragement of sound and timely publication can be achieved in industry. . . . Secrecy is unattractive in basic research, and is very seldom necessary. The communication of knowledge is a responsibility of scientists, a most important mechanism of scientific advance. Any research institution which draws value from the work of others through their publications has in some degree an obligation to return value in kind.

Researchers who harbor any of these doubts or fears will, even if they agree to write up the work, may resist having it exposed to reviewers, and the laboratory head will have to determine the validity of the objection. The competence and record of the author are most significant. A highly respected scientist with a solid reputation and an accumulation of excellent work may be given a good deal of leeway in publishing without review, or even in delaying revelation of the work. A less-experienced and untried scientist might be pressed for more justification to make certain that the cloak of confidentiality is not allowed to disguise lack of progress or flawed work.

The difficulty posed by rivalry usually occurs in fast-breaking, highly competitive fields in which the most qualified reviewers are others working in the same areas, sometimes on the same problems. The laboratory head who is sensitive to this fact and aware of the possible dangers will be able to encourage the researcher to forge ahead with confidence that his interests are being protected.

On those rare occasions when competition requires that a paper be rushed into print without review, the laboratory leader must recognize that it is a serious step that may stake the reputation of the institution on the competence and integrity of one scientist. Obviously, therefore, the calibre of the scientist is the major element in the decision.

There will probably always be some scientists who object to a review system, but many—perhaps most—reputable researchers routinely ask colleagues in the same field to review their papers before submitting them for publication. An internal review procedure is only a systematic way of providing this service for everybody, and a good system will go far toward strengthening the scientific output of the laboratory.

EDITORIAL ASSISTANCE

Scientists must not only be encouraged to write up (or prepare in proper form) their experimental work for dissemination, but the well-managed laboratory provides them with appropriate assistance to make it possible. The ideal is an editorial staff with the facilities to edit, illustrate, type, and submit articles and books to publishers; to monitor the submission until publication; to maintain records of all laboratory publications and mail reprints. The editorial office is also an excellent central point for circulation of manuscripts to internal reviewers and monitoring their progress. As soon as the review is complete and all necessary approvals given, the paper can be placed into editorial production. In this system, the whereabouts of a paper are always known, and one that gets "buried" on someone's desk is easily traced.

Highly productive laboratories often publish an annual report or compilation of significant research reported during the year. These publications are useful to science writers, researchers at other institutions, and in recruiting staff researchers. They are a source of pride to those whose work is selected for inclusion and may enhance the prestige of the laboratory within the scientific community. The preparation of such reports necessitates the skills of editors, graphic artists, photographers, illustrators, and typists—such skills as are found in the staff of an editorial department.

It takes a sizable budget, of course, to support an editorial department with this level of capability, but from the very beginning of a laboratory, the research staff should have some editorial assistance in preparing papers. Scientific editors are a rare breed and the qualifications, although very specific on some points, may vary greatly, depending upon the major research fields of the institution. Thorough knowledge of grammar, sentence structure, and composition is essential, of course, but beyond that the qualifications that have characterized excellent scientific editors have varied so widely that it is impossible to generalize. A journalist with some science education, a trade-book editor who drifted into working on scientific material, a scientist who discovered that his talent and liking for science-writing exceeded those for research—all have become fine scientific editors.

The advent of the word processor has made editing easier and considerably faster, but no machine has been invented that teaches scientists to write well or that teaches editors how to deal with the idiosyncrasies of authors. And we are still a long way, or some distance at least, from the possibilities envisioned by Eugene Garfield more than ten years ago:

> The author will sit at a computer terminal to type a new (paper) while a variety of software systems deal with the routine problems of manuscript preparation. The final draft—with all corrections inserted and bibliography automatically verified—will be transmitted by telephone lines, aided by satellites, to the journal's editorial office. There, the editor will scan it and by matching profiles, come up with the three best referees for the paper. A switching system will then transmit the manuscript to the terminals of the referees who will read it on their display screen or have it printed out for more casual reading. Referees' comments and author revisions will be transmitted back and forth through the editor until an acceptable draft is completed.

The editor may then use computer typesetting to create a highly readable, error-free copy from which a photo-offset negative is automatically generated. Or the final approved manuscript may be transmitted directly to all readers who have expressed an interest in the subject and on request to others. I can even visualize a day when scientists will hear "published" papers through voice synthesizers as they drive to work.[3]

The relationship between author and editor is always sensitive, especially when the author is a scientist. It requires patience and tact on both sides, at least until the editor has acquired some familiarity with the subject matter. There are scientists who will contest every tiny suggestion, and others who are totally indifferent. Fortunately, most fall in the middle ground. They may be mildly offended at certain changes, but if an improved manuscript is produced by the process, they accept it. It takes time for editors to build confidence in their ability because it takes a substantial number of successful manuscripts to prove that the quality of scientific writing emanating from a laboratory has improved. But it can be done by selecting the right editors and giving them encouragement and support.

AUTHORSHIP OF SCIENTIFIC PAPERS

Scientific recognition is of paramount importance to researchers, and for projects yielding patentable results or those having the potential to produce royalties, income as well as recognition may be at stake. It is impossible to establish guidelines that will cover every individual situation, but you must have criteria for settling these issues and a procedure for applying the criteria.

Any manuscript describing work done by more than one person will raise such questions as: Which names are to be included? In what order? and, particularly, Whose name appears first? The convention in straightforward uncontroversial cases, is that the senior author's name goes first, followed by others in alphabetical order. This is a reasonable form, since most citations list only the name of the senior (first) author and relegate the coauthors to the et al. category. Some well-established scientists generously will list junior associates as senior authors so that younger colleagues will be recognized for good work. But to balance these, there are some laboratory group leaders and senior scientists who insist that their names appear on any paper emerging from the laboratory, whether or not they actually participated in the work. The head of the laboratory must be keenly sensitive to this practice and be prepared to resolve complaints from the young researchers who may feel abused by their superiors on this point.

A particularly sensitive issue in some laboratories concerns the question of including the names of research assistants on publications. Because this is a matter that frequently gives rise to conflict in laboratories, members of a multidisciplinary group at one government research center got together and drew up guidelines regarding credit for published work that seem to be fair and worth noting:

> Technicians and research assistants may appear as coauthors only if their contributions clearly and substantially exceed those expected as a normal function of the position; only if they contributed to the conceptual development of the work; and only if they understand the paper well enough to present it at scientific meetings.[4]

It was their consensus that footnotes or acknowledgments were appropriate for contributions that were less than these specified.

Recent scandals involving fraudulent listings of coauthors have caused some scientists to become overly cautious—they are actually reluctant to have their names appear on publications. Therefore, it may be necessary occasionally for the head of the laboratory to insist that all those who participated in a particular study or project be included in the listing of authors.

COPYRIGHT

Once a scientific article, a graphic work, or even a computer program has been produced, it is protected by copyright. It is important for scientists to understand how the copyright system works, both to protect their own creations and, when using the work of others, to observe the appropriate procedures to avoid infringement.

Revisions to the copyright law of the United States that became effective on 1 January 1978 were the first general revisions made since 1909. The new law, enacted in 1976, made some significant changes, but there are still many controversial issues between creators and users of copyrighted material. A system rooted in the traditions of the printing press is not versatile enough to deal with dilemmas posed by modern electronics and communications technologies.

Writing in *Scholarly Communication*, issue of Winter 1986, Dennis Drabelle noted:

> By the time Congress was revising the copyright law in the 1970s, the publishers of scholarly, scientific, and technical journals—from which the bulk of educational and library photocopying is done—were worried and angry. They saw copying machines installed in nearly every college library, sometimes several on one floor, and commercial copy shops proliferating at campus borders. They strongly suspected that through a combination of interlibrary loan and photocopying one subscription to a journal was doing the "work" of many. They complained of cancelled subscriptions and declining revenues. Librarians, teachers, and students, on the other hand, argued that inexpensive photocopying was a tremendous pedagogical boon that should not be hampered by red tape.[5]

Congress has tried to assuage the animosity on both sides by the so-called fairness doctrine, which permits limited copying without royalty payment for internal use. The general provisions of the fairness doctrine allow library employees to make or distribute a single copy of a work if these conditions are met:

1. Reproduction or distribution must be made without any direct or indirect commercial advantage.
2. The library is open to the public, or available to researchers in specialized fields from other institutions.
3. Reproduction or distribution of the work includes a notice of copyright.

The ambiguity of the fairness doctrine led Congress to go a step further in attempting to resolve publisher-user disputes. It suggested the formation of a centralized permission and payment system, and, in 1977, the not-for-profit Copyright Clearance Center (CCC) was established in Salem, Massachusetts.

CCC acts as agent for about 1,200 publishers and grants to users rights to reproduce material for internal use. Fees are collected by CCC, and after deducting their agent's service charge, they remit the balance to the publishers. The fees to users are arrived at by an audit based on 90-day usage levels. CCC also offers lump-sum licenses to photocopy articles from thousands of copyrighted publications; General Electric Company opted for the license agreement when they registered with CCC. Other companies pay royalties by keeping track of their copying and paying on a copy-by-copy basis. Licensing is attractive to the business community because of their large volume of copying, particularly for those corporations with many branches scattered around the country.

Information on the CCC may be obtained from Copyright Clearance Center, 21 Congress Street, Salem, MA 01979.

An even simpler system may be on the way. One microfilm company is working on prototypes of a computer optical-disk vending machine, from which users can obtain copies of articles from selected journals in print or on optical disk. The cost is estimated to be between $4 to $10 per article. Machines can be placed in the library or other central location; the vending-machine company can bill the organization and pay royalties to the publisher on the basis of the machine's computer record.

The ease and speed with which material can be copied creates enough difficulty, but how are the old laws to deal with protection of works created jointly by humans and computers?

The Office of Technology Assessment of the U.S. Congress (OTA) was commissioned by the subcommittee on patent and trademark laws to study the matter of changes that might be needed to keep the patent and copyright laws abreast of complex electronics and communications technologies. The OTA report, "Intellectual Property Rights in an Age of Electronics and Information," issued in the spring of 1986, suggested that Congress might want to create an entirely new regulatory agency to fully meet its goals laid out in the Constitution—fostering science and the useful arts and encouraging the dissemination of information and knowledge to the public.

Registering Copyright

Until such time as a new system or a new agency is created, copyright registration is handled by the Register of Copyrights at the Library of Congress. The laws covering protection of intellectual property were revised by the 1976 law and currently "original works of authorship" may be in the following categories:

1. Literary works (scientists should be aware that computer programs and most "compilations" are registerable in this category)
2. Musical works, including any accompanying words
3. Dramatic works, including any accompanying music
4. Pantomimes and choreographic works
5. Pictorial, graphic, and sculptural works (maps and architectural blueprints are included in this category)
6. Motion pictures and other audiovisual works
7. Sound recordings

Under current U.S. law, no publication or registration or other action is required to secure copyright—unlike the previous law, which required either publication with the copyright notice or registration in the Copyright Office. Copyright is secured *automatically* when the work is created, and a work is "created" when it is fixed in copy or phonorecorded for the first time. "Copies" are defined as material objects from which a work can be read or visually perceived, either directly or with the aid of a machine or device (such as books, manuscripts, sheet music, film, videotape, or microfilm). There are, however, some advantages to registration, and it is a simple procedure; consequently, it is advisable to register works about which ownership questions may arise.

Registration establishes a public record of the copyright claim, necessary before any infringement suits may be filed in court. In the event of legal suits, registration makes it possible to claim statutory damages and attorney's fees; otherwise, only awards of actual damages and profits are available. To register a work, all that is necessary is to send to the Register of Copyrights, Copyright Office, Library of Congress, Washington, DC 20559, a package containing all the following items:

1. A properly completed application form
2. A nonreturnable filing fee of $10 for each application
3. A deposit of the work being registered. The number of copies varies

according to particular situations. Generally speaking, this is what is required:

a. One complete copy or phonorecord of an unpublished work
b. Two complete copies or phonorecords of the *best edition* of works first published in the United States on or after January 1, 1978; for works published prior to that time, two complete copies of the work *as first published*
c. One complete copy or phonorecord of works first published outside the United States
d. If the work is a contribution to a collective work and published after January 1, 1978, one complete copy or phonorecord of the best edition of the collective work.

Under certain circumstances, a work may be registered in unpublished form as a "collection" with one application and one fee.

Works Not Generally Eligible for Copyright

There are several categories of material that are not generally eligible for statutory copyright protection. They include:

1. Works that have not been fixed in a tangible form of expression—choreographic works that have not been notated or recorded, improvisational speeches or performances that have not been written or recorded
2. Titles, names, short phrases, and slogans; familiar symbols or designs; mere variations of typographic ornamentation, lettering, or coloring; mere listings of ingredients or contents
3. Ideas, procedures, methods, systems, processes, concepts, principles, discoveries, or devices, as distinguished from a description, explanation, or illustration
4. Works consisting *entirely* of information that is common property and containing no original authorship. For example, standard calendars, height and weight charts, tape measures and rules, and lists or tables taken from public documents or other common sources.

The Copyright Office has available for distribution upon request a descriptive brochure, "Copyright Basics," which can be ordered from Copyright Office, Library of Congress, Washington, DC 20559. Application forms for registration may be requested on the recently installed Copyright Office Forms HOTLINE, 202-287-9100, for callers who know what application forms they need. To find out which application form is appropriate for the work in question, call 202-287-8700 weekdays between 8:30 A.M. and 5:00 P.M.

REFERENCES

1. Day, R.A. *How to Write and Publish a Scientific Paper*. Philadelphia: ISI Press, 1979. p. ix.
2. Snow, C.P. *The Search*. New York: Charles Scribner's Sons, 1958. p. 273.
3. Garfield, E. Is There a Future for the Scientific Journal? *Sci-Tech News* 29:44, 1975.
4. Sinderman, C.J. *The Joy of Science: Excellence and Its Rewards*. New York: Plenum Press, 1985. p. 29.
5. Drabelle, D. Copyright and Its Constituencies: Reconciling the Interests of Scholars, Publishers, and Librarians. *Scholarly Communication*, Winter, 1986. pp. 4–7.

Chapter 8

LIBRARY

Many shall run to and fro, and knowledge shall be increased.

Dan. xii: 4.

J ohn Shaw Billings, the U.S. Army doctor who originated the *Index Medicus*, predicted a century ago that the time was coming when our libraries would "require the services of everyone in the world not engaged in writing, to catalogue and care for the annual product." We are not quite there yet, but librarians—and especially scientific librarians—have become resigned to the Sisyphean struggle to keep pace with the onrush of new information.

The publish or perish philosophy, endemic in the academic community and especially respected in the sciences, may be responsible in part for the proliferation of printed material. But the major credit (or blame) lies at the door of modern technology. Computers have simplified all the processes required to create, collect, store, and use information. Computerized systems for literature searching first appeared in the 1960s, and now, in the 1980s, optical and videodisc technology is again revolutionizing both storage and retrieval of archival information.

Faced with the flood of material and the continuing limitations of space, time, and money, librarians are challenged as never before to discriminate between the valuable, the mediocre, and those materials that have been tactfully described as "scientific chaff." Identifying truly worthy items requires the application of a "quality filter," and all good librarians devise their own systems of filtering, based on the needs and interests of their particular libraries. Lionel Bernstein of the National Library of Medicine, at a Rockefeller Foundation Bellagio conference in 1979, emphasized the need for such a process when he noted, "Very much of the scientific literature is not cited, not read, not sought, and not useful."

Librarians are overwhelmed not only with more information than ever before, but also with more choices of systems and machines for receiving, processing, and disseminating that information. Deciding which technologies best serve a particular library is a difficult task, compounded by improved systems entering the market almost daily, and often at lower cost. In 1984, a machine to read the information encoded on a CD ROM (compact disc read-only-memory) was priced at $16,000; in a year, the cost had dropped to one-tenth of that, and it is expected to fall within the next couple of years to as low as $250.

Qualifications for scientific librarians have always been high—perhaps the highest in the library profession—and many of them have degrees in specialized fields in addition to librarianship. The profession now demands an understanding of instruments, systems, and devices not even imagined in library schools merely twenty-five or thirty years ago. Librarians of the past were often booklovers primarily, who considered protection of the collection as their major function. Today's librarians, especially those in scientific institutions, understand that without preservation there can be no dissemination, but they also know that dissemination of information is what scientific libraries are all about. Good library management can be described as the assembly of high-quality current and historical materials appropriate to the work of the institution and the provision of systems for making those materials, and others from outside sources, available to the research staff in the most expeditious manner at the least possible cost.

The selection of the library director or, in small organizations, the librarian, is critical to the creation of an ideal information center within the research laboratory. It is, however, only one of three major factors. *Location*, both the site and the amount of space, and *budget* are the others. The best librarian in the world cannot adequately provide the necessary services if space and funds for acquisitions, equipment, and personnel are impossibly limiting.

THE LIBRARIAN

The question sometimes arises as to whether the library is an administrative function, since it is not engaged in research, or whether it is more properly within the academic or scientific structure. In a research laboratory the library is an integral part of the scientific program. Its main function is to serve the research staff by providing information and assistance at every step of the investigative process. The selection of the library director, or the chief librarian, is as important as the selection of a member of the research staff. Although a good librarian will not make a research institution great, the contribution of a superb, well-stocked, well-managed library to the research output can hardly be overestimated.

The scientific librarian needs certain professional qualifications, and a background in science or previous experience working with materials relating to the major fields of the institution are highly desirable. What's more, managerial ability is essential in a library of any size, but especially in a large one or one that is likely to expand. Selecting and managing the work of the staff are only part of the managerial skills needed. Also needed is the ability for, and interest in, long-range planning for use of present physical facilities and funds and for the eventual acquisition of additional resources. Since the library does not stand alone, such planning involves working with others within the institution responsible for space-saving and time-saving methods and machines, for building alterations and services, and for expeditious ordering systems. It also means actively participating in appropriate professional organizations, encouraging and meeting with community groups that have an interest in and can be helpful to the library, keeping abreast of new technological developments that have a relevance to library functions, and maintaining a liaison with jobbers and vendors in order to expedite the acquisition of library materials.

Librarians must be able to recognize which materials have priority and must be in the collection, which can be easily borrowed from nearby institutions, and which may be needed in multiple copies. They must also know what hours the library should be open in order to best meet the needs of the staff and what services should be available at certain hours. Since most libraries are always operating with a minimum of staff personnel, it requires some sleight-of-hand to make the necessary adjustments in order to satisfy the users and still maintain morale of the library staff.

Very few can match or even approach the world-famous library of the Marine Biological Laboratory at Woods Hole, Massachusetts. That library collects virtually everything in its fields—the sciences related to biology, medicine, and oceanography—and makes it all available at any hour of the day or night, 365 days a year, to anyone who comes looking. But, sad to say, a small cloud has appeared on the horizon; the building, with its back to the sea, is running out of shelf space, and discussions have begun on ways to reduce the collection's size.

There are some scientific libraries that do follow the example of the MBL Library and remain open twenty-four hours a day with only a roving guard or watchman to keep an eye on things. In many places like this, scientists can even check out material by leaving a card or note on the circulation desk. This practice is feasible in full-time research institutions with a minimum of transient users, but would probably not work in a large educational institution. Some colleges and universities have installed elaborate search-upon-exiting procedures, but others question whether the cost of such procedures is not greater than the losses would be on an honor system.

Patience, tolerance, and genuine desire to be a part of the research process are most important assets for the librarian in a research laboratory. Scientists are subjected to inordinate frustration in the normal course of their search for new knowledge, and it is vastly soothing to find diligent, cheerful, and persistent help in locating an article, a citation, or reference, or in compiling a needed bibliography. Therefore, the personality and attitude of the librarian are at least equal in importance to professional qualifications.

LOCATION

The location of the library makes a great deal of difference in how well it serves the staff. If possible, it should be adjacent to the research area. In large institutions, where the main library must be in a separate building, each laboratory building should have a branch library where the most-used journals and volumes are available for people who work in that area. Because experiments often involve waiting periods, the extra hours can be spent perusing current journals or looking up a reference. In laboratories where staff members travel in car pools, the library may become the gathering point at the end of the day and the waiting time spent reading or browsing.

The amount of space available to the library depends, to a certain extent, on its location. It is a mistake to jam the library into an area where there is no possibility of expansion. Space problems may be eventually alleviated by the new technologies for storing archival material and by computerizing catalogs and circulation processes, but a certain amount of growth is inevitable. Moreover, although microfilm, microfiche, compact discs, and other improvements that are just beginning to enter the marketplace may save shelf space, equipment required for storing and accessing material must be installed, and it may actually be a long time before the library can hope for much more than a trade-off in space.

One space-saving measure that is being adopted by some libraries is the formation of consortia. Libraries that are near enough for rapid interlibrary lending coordinate their acquisitions and avoid duplication unless it is justified by demand. The libraries of Cornell University Medical College, Memorial Sloan-Kettering Cancer Center, Payne Whitney Psychiatric Clinic, and the Hospital for Special Surgery (Orthopedic) have formed such a consortium. All four institutions are on an integrated, online Library Information System (LIS), and each librarian has access to the catalog and the automated circulation system of all four. These libraries are all in New York City within walking distance of each other, so if an item is found in another collection and is not already checked out, the user can get it in a matter of minutes.

Document Transfer

Only large cities are likely to have that many institutions in such close proximity using similar materials, but systems have now been introduced to greatly improve the transfer of documents by a photocopy method that is making the consortium idea more feasible.

Interlibrary loans have long been one of the most effective space-savers and the loan process has been greatly facilitated by computerization. Libraries may be part of a regional or national network through which they can locate items by computer in a few minutes. These systems expand the material available to the research staff far beyond the holdings of any one library and require no storage space.

It is not unusual for universities with numerous campuses to coordinate library acquisitions in order to avoid unnecessary duplication and concentrate materials in the branch where they are most needed. They may also have access to an integrated catalog to quickly locate books and documents. But they are usually located far apart and must depend upon the U.S. mail for transfers between the branches.

The transfer of documents has been considerably improved by a system that operates like a photocopier, called Telefacsimile. Images can be transmitted over telephone lines to compatible equipment where they reproduce a copy of the original document. This does not work for books, and photographs or detailed illustrations cannot be transmitted very clearly, but manuscripts of journal articles, reports, and typed data can be transmitted to colleagues at other institutions within minutes. Manuscripts of journal articles can also be sent to publishers using this system. It is important to remember that both institutions communicating with the Telefacsimile system must have compatible equipment. There are several types of telefacsimile equipment on the market, but Facsimile Group III is reported to be the most widely used both in this country and in Europe. In 1987, the cost of sending or receiving up to seven pages is $5; over seven pages, $5 plus 70 cents per page.

The speed of the transfer is a vast improvement over old-fashioned interlibrary loan methods, but it is still necessary to send books or photographs by courier or through the mails. Items sent out promptly usually arrive at most points in the United States within a week. Some places report that loan articles sent by rush mail arrive within twenty-four hours.

BUDGET

Budgeting for a library is rather like budgeting for a yacht. It is not a question of how much is needed, but of how much is available. There is no limit to the amount that can be spent; usually there is not enough.

The three main elements are operating costs including personnel, acquisitions, and capital equipment. At one time, equipment was almost negligible but that is no longer true. Computers, with their accompanying devices, copying machines, word processors, microfiche and microfilm readers, all have become standard library equipment. Electronic devices have also become essential for many research activities and for performing various administrative functions. Therefore, acquisitions or additions to electronic equipment for the entire laboratory, including the library, are often coordinated by a central office.

Large institutions may find that it is necessary to have a manager of computer services on the staff to evaluate and help decide what kinds of computers are needed and what kinds of sharing arrangements can be worked out among the various departments and functions of the organization. The University of Maryland School of Medicine has a director of all computing services, Marion Ball, whose staff includes a director of academic computing, and a database manager. The library, which Ball describes as "the most sophisticated science library in the country in terms of computerization," is under the direction of a professional librarian. There is also an assistant library director for public services who is responsible for bibliographic and information services.

This organizational structure places the planning and budgeting of all computer facilities within the institution in one central office where costs can be distributed on a sharing basis. This also means that the section of the library budget dealing with electronics must be coordinated with other budgets through the central computer services office.

The preparation of the operating and acquisition sections of the budget will be preceded as a rule by informal discussions with the research services manager and, in most places with the head of the laboratory. Operating and personnel costs will surely increase every year even if there are no additions to the staff, and the research services manager can advise on the projected increase because of rising insurance costs, utilities, fringe benefits, annual salary adjustments, and so forth. The same is true of acquisitions. Journal subscriptions, books, and other printed materials rise in price annually. The same level of acquisitions will continue to cost more each year. These basic increases must be taken into account before including any requests for funds to enlarge the holdings or add to the staff.

An exceedingly important element in the library budget, as is true for the laboratory budget in general, is the justification section when an increase is requested or when additional funds are sought to provide a new service. Such requests must be accompanied by a full, complete, and convincing justification. One of the strongest forms of justification is that based upon user demand; if the library can document requests for certain materials or requests for certain services, such documentation provides the basis for a most effective justification.

LIBRARY COMMITTEE

An effective library committee is indispensable to any scientific research institution. The best choices for membership on such committees are scientists who are at least somewhat scholarly. They should be relatively knowledgeable about historical documents in their fields and they ought to be those who keep up with the current literature. It is very important for committee members to understand how necessary it is to have a fine library to support a superior research staff and for them to be willing to give the necessary time to the work of the committee.

An ideal committee consists of representatives of the major research fields of the institution, and as with all committees, its mandate must be clear. The committee has a right to know exactly what matters are within its purview and whether they are to make final decisions, or recommendations only, on each matter. The committee may, for example, become a search committee to interview candidates to head the library and send their recommendation to the institutional head for final choice. In advising the librarian about acquisitions, committee approval may be final as long as the authorization is within the limits of the budget. Librarians need, and the good ones seek, advice from the committee on policy matters that affect library services, including hours the library remains open, with or without staff members present, the conditions for use of specialized equipment, and the availability of special services. Establishing policies regarding hours of use and availability of equipment and services can be complicated if the library users are in various categories (such as faculty, students, medical staff, part-time and visiting researchers). This is especially true for those functions that involve distribution of costs to user accounts.

June Rosenberg, research librarian at the Memorial Sloan-Kettering Cancer Center in New York, says that her library finds committees of most use in evaluating publications and making recommendations concerning acquisitions and deaccessioning. The library conducted an evaluation of the entire collection in 1986. When no member of the standing committee possesses the expertise to make recommendations in a particular area, the committee appoints ad hoc committee persons competent to do so.

Committee members, of course, make suggestions about the purchase of books or journal subscriptions, and all staff members should be encouraged to make suggestions about the collection, helping to ensure that the library does not become the fiefdom of one or two staff members who dominate and exert undue influence in the collection and management of the library.

JOURNALS

Scientific journals are the main artery through which current research information flows. Decisions about journal subscriptions and archival retention are among the most important and the most difficult librarians make. Journals proliferate at such a great rate that it is impossible to estimate accurately the number in existence at any given time; whenever a new field of science or area of special interest emerges, a new journal is not far behind. It has been estimated that the number of biomedical periodicals alone increases at an annual rate of five percent.

Cooperative acquisitions agreements by institutions in the same area have been effective in refining the journal collections. Institutional members of the Medical Library Center of New York—medical schools, large educational and research facilities, specialized libraries, hospitals, and some commercial firms—make commitments to subscribe to specific journals and cancellations are coordinated. Brett Kirkpatrick, librarian of the New York Academy of Medicine Library, says this works to the advantage of all members. It is, of course, most likely to work well in large metropolitan areas or specific locales where several institutions with similar interests are clustered, although such arrangements are possible between two, three, and four libraries within a limited geographic area.

Current Contents®, a publication of the Institute of Scientific Information (ISI), scans more than 6,500 journals every week and publishes the table of contents from approximately 1,000 journals in each of its five editions in the sciences. Scientists find this a useful reader's guide—a quick review of the tables of contents of the major journals in one's area or related areas of scientific interest. It helps to ensure that the most important articles will not be missed. Librarians cannot take much solace from *Current Contents*. The breadth of its coverage points up the enormity of the task of selecting which journals to buy when space and funds are always at a premium.

There seems to be wide variation in the philosophy of librarians about decisions to buy new journals based exclusively on prepublication promotion. Some librarians tend to look with favor on all new journals relating to the research of their institutions. If the publisher is well known and the editorial board and previews of articles look promising, they ask a member of the library board or another interested scientist for an opinion. If the recommendation is favorable, they subscribe. Others take a wait-and-see attitude.

Gilbert J. Clausman, who has been librarian at the New York University Medical Center for thirty-one years, says he never subscribes to a new journal on the basis of its advance publicity; he prefers to wait and see how it is received by those working in the field. When asked what he does about those journals edited by scientists in his own institution, he replied, "Those we get, but we do not pay for them. As soon as we learn that a member of the staff has written a book or is editing a new journal, we send a note saying we would very much like to have an autographed, complimentary copy or a complimentary subscription for the library. And we always get it."

One aid for evaluating journals, after they have been out for a relatively short time is the annual journal survey published as a special issue by *Nature* in the autumn. Experts are asked to comment on new scientific journals, based on at least one full year's publication.

There is no perfect way of evaluating the worth of a journal or a book to your particular collection. Several large libraries have conducted audits of their loan requests to determine the usefulness of their periodical collections, and indeed the results have been startling. A recent audit at the National Library of Medicine showed that 300 titles of their total of 22,000 periodicals answered 70 percent of all interlibrary loan requests. In 1985 a

Rockefeller Foundation–funded bibliometric study of the Marine Biological Laboratory library at Woods Hole, Massachusetts analyzed the activities in the library's journal collection over a typical ten-month period. Kenneth S. Warren, director of health sciences at the Rockefeller Foundation, and others suggested that the results of the study supported the idea of reducing the size of the MBL library. Of the 2,437 journals to which the library currently subscribed at the time of the study, approximately 20 percent were completely unused during the ten months. Of the 4,765 titles in the journal collection, about half were not called on even once. Almost two-thirds of all journals in the collection averaged fewer than one use per month. More than half of the journals used were current issues; more than three-quarters were dated after 1980; only 5 percent were from the pre-1961 collections. In all, about 2 percent of the total collection, just 113 titles, accounted for fully half of the total journal use. With just 250 titles, the library could have filled two-thirds of its requests. Ten percent of the total collection received 80 percent of the readership; 27 percent of the titles answered 95 percent of the requests.[1]

The findings of the Rockefeller analysis of MBL are consistent with those of the first bibliometric study done in 1956 at the British Lending Library for Science and Technology (now named the British Library Document Supply Centre). The study showed that of the 9,000 periodicals in the collection, forty sufficed to fill half the requests. By adding fifty titles, for a total of ninety, or 10 percent of the collection, the library could fill 80 percent of all requests.

If circulation counts were the only factor to consider, paring down the journal subscriptions would be simple. But the science library has the pressing obligation and concern to provide every researcher with needed information even though some material is useful to only a few members of the staff. Many journals cover narrow, highly specialized areas but are essential to the relatively few scientists engaged in those particular fields of study. Wide readership is not a reliable criterion for measuring the usefulness of a journal to a research institution.

Another method of evaluating journals is citation analysis, which was introduced by librarians in the 1920s to help evaluate journals. They assumed that frequent citation was a measure of effectiveness in the sense that heavily cited work was useful to other researchers and in promoting further work. In the beginning, the correlations had to be done by hand, but now there is ISI's *Science Citation Index® (SCI)®* and *Social Sciences Citation Index® (SSCI)®* which cover the international literature of science and index more than 600,000 items annually from every scientific discipline. Scisearch, an on-line version of SCI, available through Dialog and BRS, simplifies the analysis, but ISI goes even further in facilitating the process. Each SCI subscription includes *Journal Citation Reports® (JCR®)*, which is designed for statistical evaluation of scientific journals. JCR can be used to answer the questions: How often has a particular journal been cited? What journals have cited it? How frequently have particular journals cited it? Is it the older or newer material that is being cited? Besides aiding the librarian in collection management and journal acquisition, the JCR may help researchers in deciding where to submit work for publication.

Not all scientists agree that citation analysis is an accurate measure of quality. Some argue that the premise is faulty, that citations are used for a variety of reasons and thus quality cannot be inferred from quantity. For example, a flagrantly inferior paper may be widely criticized, and thus highly cited. They also suggest that citations from highly regarded journals should count for more than those from obscure journals. Others say that papers describing methodological techniques may lead to long lists of citations when the techniques become useful in basic research. Therefore, journals that publish methodological papers may receive many more citations than those devoted to basic research, which contribute more to science. For example, the single most highly cited paper between 1955 and 1986 reports on a method: "Protein Measurement with the Folin Phenol Reagent," by O. H. Lowry, N. J. Rosebrough, A. L. Farr, and R. J. Randall, was cited 167,154 times during this period. The next most highly cited article also reported on a new technique and received only 42,102 citations. The vast majority of articles cited in that period had fewer than 100 citations.

In established research institutions, where a basic journal collection is in place, the major problem lies in dealing with the surfeit of materials that become available in a steady flow. There are other libraries, however, in poorly funded institutions or those that are just starting up, particularly in developing countries, where the building of a substantial research library is well-nigh impossible.

In the past five years, the Rockefeller Foundation has sponsored a demonstration project to improve access to the world's medical literature, particularly in Africa, Asia, and South America. The project has developed the *Selective Medical Library on Microfiche*, which is being tested in four libraries in Indonesia, Egypt, Colombia, and Mexico. The core collection of ninety-one titles started with a nucleus of 35 journals selected from several significant lists, such as those of *The New England Journal of Medicine*, the *Abridged Index Medicus*, and the Brandon List. Based upon citation data from the nucleus of 35, a list of about 100 journals that highly cite or are highly cited by the 35 was generated; 91 of these 100 are currently available in microform. These 91 were broken down into 37 specialty clusters and the other nine will be added as they become available. The period covered for the 91 journals is the years 1982–1987.

The importance of microfiche journals for the purpose of this project is that they require less storage space, are relatively immune to climatic conditions, and are less expensive than printed journals. Microfiche can also be sent by airmail, which saves shipping costs, avoids delays in surface mail, and reduces the danger of loss.

These advantages are not equally attractive in the United States where most researchers prefer to work with materials in hard copy. Most libraries, however, are now equipped with microfilm/microfiche readers, since some material is received only in that form. Many government reports, for example, are distributed on microfilm. The main objection researchers have to the use of microfilm reader-printers is that the copies of illustrative material are not as clear as those made by photocopy directly from the printed page.

Cyril Feng, director of the Health Sciences Library at the University of Maryland–Baltimore, says they do not put journals on microfiche and have no plans to do so unless they run out of space. Jean K. Miller, director of the superb library of the University of Texas Health Science Center at Dallas, also agrees that researchers in the medical and biomedical fields prefer to work with documents in hard copy. Therefore, they have very few journals on microfilm, retaining even some of the oldest ones in hard copy. The Dallas library is particularly fortunate to have ample space, at least for the time being.

S. K. Cobeen, director of the Library of the United Engineering Center in New York (an organization of several combined engineering societies), says engineers do not differ from life scientists in this. They also prefer hard copy. However, this Rockefeller-sponsored core collection of important journals on microfiche in the international biomedical and health literature may be a valuable acquisitions guide for biomedical research libraries anywhere in the world that are just getting started or that are refining their journal collections. Additional information on the *Selective Medical Library on Microfiche* is available from Kenneth S. Warren, M.D., Director of Health Sciences, The Rockefeller Foundation, 1133 Avenue of the Americas, New York, NY 10036.

The best system of all is experience. A librarian gains insight from working with the research staff—a more precise indicator of the needs of a particular library than an objective system. Judgment takes time to acquire and the process never ends, since it is the nature of scientific research to enter new areas of investigation as soon as old questions have been answered and new ones posed. It is imperative that the librarian be sensitive to researchers' requests for materials and be willing to spare no effort in locating them. Working closely with the library committee and asking for advice and recommendations regarding acquisitions is indispensable to the provision of good library service.

ARCHIVES STORAGE

Archival storage is on the brink of a technological breakthrough that will change our mental image of what a library looks like. Instead of row upon row of tall stacks, filled with bound volumes and labeled at each end to indicate the holdings of each, we may see tiers of neat metal cabinets, each containing hundreds of tiny cassettes or plastic cards coded with information. Instead of scholars seated at long sturdy tables, poring over documents under shaded reading lamps, there will be rows of researchers bent over machines reading, playing, or printing the material from cards or discs. Books bound in soft leather, tooled with gold lettering, will be a curiosity. But, of course, that is already true. Libraries—except for those great ones with collections of rare books—have not had books bound with beautiful leather for at least a century.

Optical memory was a dream planted somewhere in the twenty-first century when the first computer databases came on-line in the early 1960s. But, thanks largely to the aggressiveness of the home-entertainment industry, CD ROMs—a term that first appeared in the popular press in 1984—are now entering almost every market for storing information. The CD ROM (compact disc, read-only-memory) was developed by Philips in the Netherlands and Sony Corporation in Japan to reproduce music. In only two years these discs have established themselves as a significant part of the home music market. The CD ROM is smaller than the videodisc, just 4¾ inches in diameter, and holds about 600 million bytes of information, compared to 1,000 bytes on the videodisc. It cannot, however, store images as efficiently as the larger disc.

The CD's compact size, large storage capacity, rapid access ability, and low cost make it an attractive medium, especially for libraries. Their primary use at the moment is for storing bibliographic databases, especially those that must be updated regularly. They are so inexpensive that outdated copies can be discarded.

At the world's largest computer-industry trade show, the Spring Comdex, held in Atlanta in 1986, CD ROMs were on display that had been adapted to hold encyclopedias and their indexes. The entire Grolier encyclopedia, plus an index nine times its size, is now available on a single disc. By 1986, a number of libraries had already installed the CD ROM bibliographic system.

Another form of optical memory, called optical digital data disk (ODDD), stores up to 4 billion bytes of information in a form that is directly readable by computers. The disc serves as its own master, can be easily duplicated, and does not need to be sent out for mastering. The discs are stored in players that have been likened to old Wurlitzer jukeboxes. Instead of filling out a stack request, an accession number is keyed in and the information is sent within six seconds to an electronic buffer beside the requestor's terminal. Then, on a high-resolution screen, the book or journal can be read page by page as slowly as the requestor likes, or the document can be printed on a high-speed printer.

The Library of Congress, which is running out of storage space and has a large percentage of its collection disintegrating on acid-based paper, embarked upon an ODDD pilot project. One part of the project was to optically scan the 1984 and 1985 editions of seventy-three journals and store them on discs as facsimile pages. Because each disc is its own master, eventually the library could duplicate the discs and send them anywhere in the world. This raised questions about copyright protection; under the Copyright Act of 1980, input to a computer is considered a copy. The Library of Congress created an advisory committee made up of publishers and librarians to work out a policy statement concerning the use of the optical-memory technology for copying and distributing copyrighted material (see chapter 7, "Publication of Research Results/Copyright.")

Unlike magnetic media, all forms of optical memory are very durable—you cannot ruin them with a paper clip or by writing on them. Since optical memory is still in a read-only medium, it does not yet replace magnetic tape and discs. But at the Comdex show in Atlanta, companies were showing systems that could record CDs as well as play them back. There is some uncertainty as to how soon they will be technologically reliable, economically feasible, and generally available, but this is the system that is likely to bring about the greatest change in information storage, particularly for libraries. It may be sooner than we think. After all, it was scarcely more than an idea just a little over 20 years ago, and the pressure for solutions to diminishing space and deterioration of historic documents printed on acid-based paper may be effective in bringing these systems to the marketplace in the very near future.

Hand in hand with determining the methods of archival storage goes the question of what is to be preserved and in what form. In this, as in acquisitions management, the experience of a librarian in a particular institution is most valuable, and also, as with acquisitions decisions, the advice and guidance of members of the research staff—the library committee and others—is indispensable.

Immediately following the issues of what to store and by what process comes the most important issue, that is, making the material available to the researcher quickly and conveniently. If one has to stand in line to check out discs and then wait for an available reader or player, it may be more expeditious to go into the stacks for a bound volume. It is worth noting, for those libraries that prefer to bind back issues of journals, that poorly labeled stacks can be very frustrating. It is surprising to see that some libraries give only the name of the journals with no indication of the dates of issues contained in each stack. If one is looking for a particular number of a weekly journal of ten or more years ago, it may be necessary to browse through several stacks before getting to the right year. This is one of the small services requiring very little extra effort on the part of the library staff that can mark a library as "user-oriented" and not "staff-oriented." This applies equally to microfilm or CDs that may be made available in open stacks.

User-oriented libraries know that photocopying facilities are essential for researchers working with hard-copy materials. They usually prefer to photocopy documents and take them away to study, even if they read them in the library; photocopies allow them to make notes on the document itself and to save it for future reference. Copying machines should be located in areas where the traffic they generate does not interfere with other library use but near enough, especially to those shelves with the heaviest tomes, to be convenient. For information about photocopying copyrighted publications, see chapter 7, "Publication of Research Results/Copyright."

LITERATURE SEARCHING

Libraries in research institutions might easily be called "literature supply rooms." "Supply" implies both storage and distribution, and it also assumes expertise in locating what is needed. Information holdings of a research library are only a part—perhaps the smallest part—of the supply of information it can make available.

Computers have made it possible for libraries to form consortia and other links that make the facilities of each library in a group available to all others. Group members may be close together in large cities, spread out over several states, or even national and international in range. It is not an exaggeration to say that almost any information that is known anywhere in the world today can be made available to scientists working at an isolated archaeological dig in Anatolia. It is only a matter of having the right equipment and pushing the right buttons. One example of cooperating libraries is the South Central States group based at the University of Texas Health Science Center at Dallas Library called TALON, for Texas, Arkansas, Louisiana, Oklahoma, and New Mexico.

On-line searching is a means of gaining access to information stored electronically in a computer database. Just as researchers would approach a card catalog or an index with a set of terms that describe the subject in which they are interested, they can send those terms to a computer database, via their terminals and a special command language used to relay instructions to the computer, and can receive a list of documents relevant to their subject of interest.

On-line searching has the advantages of speed, precision, and currency. Its relatively low cost has made it of use generally in libraries everywhere in the developed world. Charges are usually based on the time actually spent on-line directly and may range from $15 to $300 per hour, often plus a small fee for every retrieved item. A typical search has been estimated at ten to fifteen minutes, but the total time depends upon the complexity of the search and also on the skill of the searcher. It is possible to search through several million records for about $25. Martha Williams, of the University of Illinois at Champagne-Urbana, says that in ten or fifteen minutes one may search records equivalent to the Library of Congress catalog, twenty years of *Chemical Abstracts*, or ten years of *The New York Times* and dozens of other newspapers.[2]

Computer searching began in the 1960s and databases generally include material since 1965. Gradually, however, some databases are being expanded to include earlier information. The Chemical Abstracts Service (CAS) Registry and Nomenclature database provides registration number, CA nomenclature, and synonyms for some seven million chemical substances, adding about 345,000 each year since 1965. Recently, it began to register substances cited in *Chemical Abstracts* from 1920 to 1965, adding around 600,000 annually.

The equipment necessary to perform on-line searches includes a computer terminal or microcomputer; a telephone line; and a modem, a device that modulates/demodulates signals communicated between your terminal and the computer you are accessing. (It is also the device for which the National Library of Medicine named a feature in its *Technical Bulletin*, which publishes information supplied by users of the MEDLARS database—it is called "Easy as Pie a la Modem.")

The number of terminals needed, the required storage capacity of the microcomputer, and the connections with other computers in the same institution are all factors that depend upon the size and needs of each particular library and the computer facilities available or planned within the institution where they are located.

Expert advice and assistance is essential in selecting computer equipment. This is a rapidly developing field and only a specialist can weigh, with any degree of accuracy, projected serviceability against current cost.

On-line search service organizations are vendors who sell access to databases produced by others. There are currently approximately 370 vendors who distribute a total of 2,805 databases, of which 1,092 are in science, technology, and medicine.[3]

Researchers in the health sciences have access to the National Library of Medicine's (NLM) Medical Literature Analysis and Retrieval System (MEDLARS), which contains six million references to journal articles and books in the health sciences published after 1965. Most of the references have been published in *Index Medicus* or in other printed NLM indexes and bibliographies. Arrangement for access to MEDLARS can be made directly with NLM by an agreement that includes training in searching the on-line databases of the MEDLARS system. Databases of MEDLARS may be accessed four ways: direct dial to NLM, or through the TELENET, TYMNET, or UNINET networks. Nineteen databases are currently available, the largest and most frequently used being Medline, which covers 3,000 journals published in the United States and other countries, and contains 800,000 references to biomedical journal articles published in the current and three preceding years. Coverage of previous periods (back to 1966) is provided by backfiles searchable on-line that total some 3,500,000 references.

Medline is also available through the two major database vendors, Dialog and BRS (Bibliographic Retrieval Services). However, access to the MEDLARS system includes, in addition to Medline, eighteen other databases that are not all available through commercial vendors. The

NLM system is less expensive since it is government supported and is available virtually twenty-four hours a day, seven days a week. Also, NLM databases are not copyrighted and therefore can be copied and retained. The Medline database accessible on other systems is the same material, but organized differently and therefore it may be useful to have access to the others as well as the NLM system.

Another MEDLARS database, Catline (Catalog Online) gives medical librarians immediate access to authoritative cataloging information on about 500,000 books and serials cataloged at NLM, thus reducing the need for librarians to do original cataloging. This is also a useful guide for ordering books and journals and providing reference and interlibrary loan services. Complete information on MEDLARS is available from MEDLARS Management Section, National Library of Medicine, 8600 Rockville Pike, Bethesda, MD 20894.

Dialog is a comprehensive information resource with more than 250 databases in engineering, chemistry, biosciences, research, patents, law, medicine, education, and business management. Complete information on Dialog databases is available from Dialog Information Services, Inc., 3460 Hillview Avenue, Palo Alto, CA 94304.

BRS is an integral part of the Information Technology Group (ITC), six companies that provide information to the engineering, medical, library, financial, government, and retail markets in the form of hard-copy publications, microform services, and on-line computer access. It is one of the largest vendors of on-line databases in the world and it also supplies, as a separate product, a system for conveniently storing and rapidly retrieving information called BRS/SEARCH, for those who wish to create and search their own electronic files.

BRS markets databases in the fields of medicine, biosciences, science, health, social sciences, politics, and business, and related interdisciplinary areas. Late in 1984, five medical journals joined in a venture to make their articles available through the BRS/Saunders database, Colleague. Articles from *The New England Journal of Medicine, Annals of Internal Medicine, British Medical Journal, The Lancet,* and *The Medical Letter* are available through this database, in some cases at the same time as the paper-and-ink versions. Information about Colleague and all BRS services is available from BRS Information Technologies, 1200 Route 7, Latham, NY 12110.

STN International is a cooperative on-line service based on a telecommunications link between the computers of the Chemical Abstracts Service in the United States and the computers of the Information Center for Energy, Physics, Mathematics in Germany. The link makes the Chemical Abstracts Service (CAS) on-line databases accessible to European users through the German node of the network and the PHYS (Physics Briefs) database of the Information Center for Energy, Physics, Mathematics accessible to North American users through the United States node. In the United States, detailed information on STN may be requested from STN

International, 2540 Olentangy River Road, Box 02228, Columbus, OH 43202.

Databases made available through vendors are produced by professional organizations. For example, the following databases are available through both Dialog and BRS, organized and indexed in different ways:

SPIN, produced by the American Institute of Physics, covers research in physics, astronomy, and geophysics from articles in major American and Russian physics and astronomy journals.

Scisearch is a multidisciplinary index to scientific and technical literature from over 4,500 journals in the natural, physical, and biomedical sciences, produced by the Institute of Scientific Information (ISI).

CA Search, produced by the American Chemical Society's Chemical Abstracts Service, covers the chemical literature from 1967 to the present and corresponds to references in *Chemical Abstracts.* It is available to European users through STN International.

Biosis (Biosciences Information Service) is a broad-based information service covering the field of biological and biomedical sciences. It publishes abstracts, content summaries, and indexes of journal articles, meetings, reviews, reports, and books from the world's life-sciences literature, and makes this material available through the Biosis database.

Even with access to numerous databases through several systems, the searcher sometimes has to be ingenious, intuitive, and, above all, persistent in order to locate a document. For example, Mary Esther Gaulden, a professor and research scientist at the University of Texas Health Science Center at Dallas, wanted to find a paper on chromosome effects of thalidomide written by a scientist in Iran in 1971. The *Chemical Abstracts* reference indicated it was published in *The Quarterly Bulletin of the Faculty of Science of Teheran University.* As this publication was not in her university's library, her interlibrary loan department initiated a search for the article from libraries in the United States. After searching the files of numerous libraries that have comprehensive holdings of the medical and scientific literature without success, Gaulden wrote to Teheran University and to the British Library Document Supply Centre, famous for being able to produce almost any piece of known scientific literature. The Iranian university did not reply, and the British library could not provide the article. Finally, in desperation, she went to the most recent annual *Science Citation Index's Source Index,* an alphabetically arranged author index that gives, among other information, the full addresses of authors who have published in the year of the *Index.* It was a long shot, based on the outside hope that the author had published recently and that his name was included in the *Source Index.* He was, in fact, listed with a California address. Gaulden telephoned him immediately and in several days received a reprint of the article, thus ending a worldwide search.

Computers have not made printed journals, abstracts, and other scientific and technical literature obsolete. Questions about the future of scientific journals have been raised for several years. In 1975, Eugene Garfield concluded (accurately, so far) that, "despite the sci-fi trappings, . . . the journal will still be a journal, filling the same functions it does today."[4]

Aside from the fact that most databases date back to only 1965, most of them are merely guides to the location of full-length articles, although some services are beginning to offer the entire articles. Generally, the computer search for relevant citations is only the first step in obtaining the needed material. Libraries must still collect journals and other primary sources of scientific information as well as secondary sources such as *Physics Abstracts*, *Biological Abstracts*, *Chemical Abstracts*, and others.

Libraries with limited archives, particularly those in isolated places without access to a major library, may find ISI's "The Genuine Article" service helpful. It provides original tear sheets (articles actually removed from the journals in which they appeared) from more than 7,200 journals for the current and four preceding years. When the original tear sheet cannot be sent, ISI provides a good-quality photocopy. For further information, write to Customer Services, ISI (Institute for Scientific Information), 3501 Market Street, Philadelphia, PA 19104).

Publishers of journals and other scientific literature are among the major producers of computer databases. Information Services for the Physics and Engineering Communities (INSPEC) of the Institution of Electrical Engineers in England publishes a variety of abstracts journals, among them *Physics Abstracts*, *Electrical and Electronics Abstracts*, *Computer and Control Abstracts*, and *Key Abstracts*—a series of eight monthly journals in several technical fields. They also produce the database INSPEC from their publications. INSPEC-1 provides the complete document record including the abstract and is available in four separate sections that correspond to the four abstracts journals, or in any combination of sections. INSPEC-2 is designed to provide current awareness and appears in advance of INSPEC-1, giving only the bibliographic reference, classification codes, and subject indexing.

The National Federation of Abstracting and Information Services (NFAIS) is a group of member organizations that publishes abstracting and indexing and related services in all subject areas, in print and computer-readable form. The federation holds annual meetings on subjects of current interest to its members. The theme of the 1986 meeting was "Innovations in Information Delivery." The meeting featured a preconference seminar, "Alternatives to Online: New Delivery Formats," at which talks included "CD ROM: Prospects and Realities" and "Laser Card: Two Million Characters of Text on a Credit Card." The federation publishes a monthly newsletter containing items of current interest in the field. For

---

ON-LINE DATABASE REFERENCES

An Introduction to Online Searching. Contributions in Librarianship and Information Science, no. 50. By Tze-chung Li. Westport, CT, Greenwood, 1984. 289 pp. Bibliography, index. $27.95.
This book features a critical bibliography of sources pertaining to on-line searching.

Online Searching Technique and Management. Edited by James J. Maloney. Chicago, American Library Association, 1983. 192 pp. Bibliography. No index. $25.
The first section is an introduction to on-line searching; the second an examination of such management issues as the policy manual, record-keeping, budgets and funding, training, and publicity.

Directory on Online Databases. Santa Monica, CA, Caudra Associates. 5 indexes. $95/year.
A regularly updated, comprehensive directory of print databases, published quarterly. Subscription includes two complete master directories and two supplements annually. The 1985 edition lists approximately 2,760 databases available throughout the world.

Computer-Readable Databases: A Directory and Data Sourcebook. 2 vols. Edited by Martha E. Williams. Chicago: American Library Association and Amsterdam: Elsevier, 1985. Pagination varies. 4 indexes to complete set in each volume, 2-vol. set $157.50.

Libraries and Information in the Electronic Age. Edited by Hendrik Edelman. Philadelphia, ISI Press, 1986.
A compilation of twelve lectures presented in the early 1980s by professional librarians and information scientists on the "information age" and its effect on libraries and information science.

---

information about annual meetings, the newsletter, and other publications, write to National Federation of Abstracting and Information Services, 112 South Sixteenth Street, Philadelphia, PA 19139.

The American Federation of Information Processing Societies, Inc. (AFIPS) also holds annual meetings on information processing and publishes the proceedings. For information on their meetings and publications, write to AFIPS Press, 1899 Preston White Drive, Reston, VA 22091.

Librarians continue to look toward a future in which automation will play a more important role, but the millennium has not arrived. The new technologies emphasize the necessity for both long- and short-range planning, taking into account fiscal realities, cost-benefit ratios, and the problems and limitations imposed by the almost-daily appearance of new and better equipment. This trend toward a more electronically oriented library prompts library schools to continuously modify course content to reflect what is happening. And managers of scientific institutions may find that even the best librarians need training in new technologies occasionally and should be prepared to assist them in obtaining it.

REFERENCES

1. Wolff, A. A Cold Eye on Mediocrity. *RF Illustrated: An Occasional Report on the Work of the Rockefeller Foundation.* March, 1986.
2. Williams, M.E. Electronic Databases. *Science* 228 (4698): 445–456, 1985.
3. Williams, M.E. (ed.) *Computer-Readable Databases: A Directory and Data Sourcebook.* Chicago: American Library Association, 1985.
4. Garfield, E. Is There a Future for the Scientific Journal? *Sci-Tech News* 29(2):44, 1975.

Chapter 9

INVENTIONS AND PATENTS

There are science and the applications of science, bound together as the fruit to the tree which bears it.

Louis Pasteur—*Revue Scientifique*

We exist in an industrial society, and scientists have an obligation to see that any useful product or process developed or improved by research or study is made available to benefit humanity. The patent system is one of the most effective ways of assuring that discoveries made in the laboratory—whether academic, industrial, or government—are transferred to the marketplace. In advance of the emergence of a patentable discovery with commercial potential, research administrators have a responsibility to establish patent policies that define the rights and obligations of both the inventor and the institution. Most organizations execute agreements concerning inventions as a part of their employment procedures, outlining their policies regarding rights to inventions, income-sharing, and so on. This is a good idea. Everyone understands right from the beginning what to expect if (and when) a patentable process or product results.

The size of the research laboratory or institution does not matter. The number of patents issued in the United States exceeds 4 million, and except for a very few, like television and transistors, the great ones have resulted from the work of one individual or of very small laboratories in colleges and universities; for example, laser beam, radar, and photocopying. If there is no clear-cut policy, and particularly if a discovery is made in the course of research that is partially or entirely funded by an outside sponsor, the outcome can be very unfortunate.

Some bitter and some heartbreaking experiences have clouded the lives of researchers whose discoveries came to have great commercial value, and from which they gained nothing. Enrico Fermi, who discovered nuclear fission in the 1930s, planned to establish a fund for Italians to study in the United States with the money he believed he and his collaborators would receive—perhaps some $10 million—as compensation by the government for the use of his patent in the manufacture of isotopes. After long and costly litigation, a settlement was made, but it was so small it barely covered the lawyers' fees.

A patent is a property right granted by a sovereign nation that gives the holder the exclusive right to exclude others from the manufacture, use, and sale of an invention in that country for a period of years. As a property, it may be sold or assigned, pledged, mortgaged, leased (licensed), willed, or donated, and may be the subject of contracts and other agreements. The owner of the patent may produce the invention commercially or may license others to do so.

The United States Patent Office examines applications for novelty, utility, and nonobviousness, and it is the applicant's responsibility to establish that these elements are present before the patent is issued.

The statutory eligibility for patenting in the United States includes "a new or useful process, machine, manufacture, or composition-of-matter, or any new and useful improvement thereof." This was extended by a Supreme Court ruling in June 1980 to include life forms resulting from genetic engineering. Also included are new varieties of asexually produced plants, other than tuber-propagated plants or plants found in an uncultivated state.

The laws are very specific about what *cannot* be patented in the United States: theories, ideas, plans of action, results, methods of doing business, discoveries of laws of nature or scientific principles, things immoral or injurious to health and the good of society, and works eligible for protection under the copyright laws.

Premature disclosure of an invention by publication in a scientific or technical journal or by public use may place an invention in the public domain and render it ineligible for a patent. In the United States, a patent may be obtained if a patent application is filed within one year after the invention is disclosed through publication, sale, or public use. In many foreign countries, a patent cannot be obtained if there has been any disclosure—even oral—of the invention to the public prior to the filing of a patent application. However, under an international convention, a patent application in the United States generally will preserve for one year the right to file patent applications abroad, even though there has been publication of the invention after the filing of the U.S. patent application and before the foreign application is filed.

Although the official disclosure date is the date of publication, patentable information is actually disclosed when an article is submitted to the publisher. It is therefore important that records be kept of the dates manuscripts are received by publishers or editors and the dates they are accepted for publication. Some scientific journals, such as *Science* and *Nature*, publish the dates of receipt and acceptance with the article. Those dates become exceedingly important in establishing priority in connection with patent negotiations.

INSTITUTIONAL POLICIES

Most research institutions now have well-articulated patent policies, but there is still potential for conflict between an institution and its employees when it comes to identifying all those who contributed to the invention and to determining the distribution of income.

George H. Hitchings, who directed research at the American division of Burroughs-Wellcome until his retirement in 1975, felt that it was impossible to distribute invention credit and income fairly. Since all scientific work builds upon, and makes use of, the work of others, he says, "Patent agreements create more problems than they solve. They encourage secrecy; and some research directors want their name on everything. So many people are involved from bench to product it is unfair to single out one for royalty sharing. There are other ways to reward scientists." Hitchings, as others, believes that the institution should retain all patent income, a not-uncommon policy in many industrial laboratories.

When research funded by an outside sponsor—government, foundation, or corporation—leads to a commercial product, the credit and distribution problem can become very complicated indeed. A great deal of research at colleges and universities today is funded by corporations and each organization involved needs to have a well-defined patent policy when it comes to grant-negotiating time. If the policies conflict, it may be possible to arrive at an arrangement through negotiation that is satisfactory to both parties. If the conflict is so great that agreement seems impossible, it is best to forego the grant. Disappointing as that may be, it is preferable to risking a time-consuming controversy, perhaps even a legal battle, that can only be detrimental to all.

After negotiations for a research grant have been successfully concluded between two organizations, incoming grants and contracts should be continuously monitored to see that no changes in policy have occurred since the original understanding. Foundation and corporate policies can change frequently. Policies at educational institutions and in the federal government can change with even greater frequency.

Even when there is a well-defined policy, if it is not properly administered, the results can be tragic, or even ludicrous. Consider the "Gator-Ade" case. This commercial product resulted from experiments conducted at a Florida university, under a federal grant. The funds covered purchase of materials and equipment and payment to human subjects, namely, the university's football team. The investigator, as soon as he discovered that he had come up with an effective energizer, filed a disclosure statement with the appropriate university office. It got buried on somebody's desk and nothing happened. After waiting for what seemed to him a reasonable time, the investigator proceeded to set up a private foundation and in partnership with the Stokely–Van Camp Company obtained a patent and entered into a licensing agreement under which Stokely–Van Camp had exclusive production and marketing rights and shared income with the foundation. When the information became known, the Department of Health, Education, and Welfare (now Health and Human Services) and the university—in a previously unheard-of action—brought suit against the foundation and the licensee. The case was settled out of court, amicably, by an arrangement whereby income was to be shared by the university, the Stokely–Van Camp company, and the foundation (the investigator). This is an amusing story because it has a happy ending, but if the university had submitted the disclosure statement and exercised its option to apply for a patent (permitted at that time by DHEW), a wasteful legal battle and a lot of confusion might have been averted.

A "Suggested Outline for Disclosure of Invention," used by Battelle Development Corporation is shown on page 144. (Battelle has used it for years but is not certain of its origin.) The form can be revised to serve the purpose of each research laboratory; its significance is that it indicates the type of information that must be included in a disclosure statement.

U.S. FEDERAL GOVERNMENT POLICIES

United States government policies regarding inventions resulting from research supported by federal funds have been updated and clarified since the Fermi and Gator-Ade cases. Public Law 96-517, enacted by Congress and signed by President Carter on December 12, 1980, which became effective on July 1, 1981, gives nonprofit institutions and small businesses the option to retain title to inventions resulting from work performed using funds from *any* federal agency (except the Tennessee Valley Authority). It also provides that:

1. All inventions must be reported to the funding agency within sixty days, after which the institution has a stipulated time to take title to the invention.
2. The government will retain a royalty-free license to the inventions and, if it deems necessary—if inventions are not brought to commercialization after a specified number of years—may exercise so-called march-in rights.
3. Patent holders may negotiate exclusive licenses.
4. The individual inventor must share in the royalties.

OMB Circular A-124 includes a requirement that institutions using federal funds for research sign written agreements with all employees (except clerical and nontechnical employees) concerning invention patent rights.

Subsequent legislation, Public Law 98–620, November 1984 and Public Law 98–622, January 1985, further amended and refined the statutes governing inventions at nonprofit institutions and small businesses using government funds. All changes and proposed changes in patent laws are routinely published in the *Federal Register*. A list of patent-information sources is provided on page 147.

SUGGESTED OUTLINE FOR DISCLOSURE OF INVENTION

INVENTION DISCLOSURE:	(Title)
INVENTOR(S):	(Print full names)
SUBMITTED BY:	(Identify name of University, College, Company, Laboratory, Institution, or indicate if Individual Inventor)
OBJECTIVE:	(General purpose of invention)
STATE OF ART:	
Current:	(Describe present method(s), if any, of performing the function of the invention; list references if possible.)
Disadvantages:	(Disadvantages of current methods)
INVENTION DESCRIPTION:	(Construction of invention showing changes, additions, and improvements over old methods. Where applicable, illustrate invention with sketches, drawing or photographs in which the parts referred to are identified by reference number or by name. Give details of operation. You are encouraged to reference and attach any write-ups of invention used for other purposes for description.)

(If the invention relates to composition of matter, include the following additional points in the disclosure:

    a. Show the general properties required for each class of materials used. If possible, list at least three examples in each class. Explain the method of preparing any new material for which a method of preparation is not already known.

    b. Set forth *proportions* of materials, and conditions, expressed in the form of the widest reasonable ranges that will work. Also, mention narrower limits within these ranges that will provide optimum results. State the disadvantages of using proportions or conditions outside the ranges selected.

    c. Give *specific examples* of practice of the invention, in various modifications, and with the preferred proportions and conditions. The examples should illustrate diverse conditions under which the invention may be practiced.)

Novel Aspects:	(Describe any *surprising results* that would not have been forecast by an ''expert in the art''. Explain the surprising results, if possible. Emphasize any results that are contrary to what was expected.)
Principles:	(Explain the operating principles of the invention.)
Advantages:	(Over what has been done previously)
Disadvantages:	(Biggest question or weakness)
Alternate Methods:	(Of construction or operation of invention)
PUBLIC DISCLOSURE OF INVENTION:	(Thesis, Abstract, Speech, Article, or Other. Include copy and date if possible. If a speech, give date and circumstances.)
COMMERCIAL INTEREST:	(Companies that have expressed an interest, or should be interested in the invention.)
COMMITMENTS:	(If supported by government contract, identify grant or contract and rights available. If individual, identify other assignees.)
WITNESSES:	(After the disclosure is prepared, it should be signed at the end by each inventor, as indicated below. The disclosure should then be read and witnessed by at least one other person, as indicated below.)

INVENTOR _____   _____

              First Name   Middle Initial   Last Name   Date

Disclosed to and understood by the undersigned on the dates indicated:

WITNESS _____ Date_____

PATENT MANAGEMENT

The formulation of an institutional patent policy is one of those managerial functions that can benefit from the collective wisdom of several people representing various interests—scientists, lawyers, ethicists, administrators. In addition to recommending and establishing policy, a committee is needed to oversee the administration of the policy, to adjudicate disputes, and to determine which inventions are sufficiently promising for the institution to apply for a patent. When it comes to the management of patents and negotiation of licenses, only a fairly large institution can establish a department qualified to handle those matters.

Patent management is a highly skilled field requiring the services of experts in a number of professional areas. In many cases, research institutions conclude that the volume of patentable results likely to emerge from their programs does not justify the expense of setting up an in-house patent-management department. Some nonprofit institutions, primarily colleges and universities, have established institution-affiliated foundations to handle patents and licensing. In-house management and foundation management allow the inventor to work closely with those promoting the invention, and these methods ensure that maximum income will accrue to the institution and the inventor. But both foundations and in-house management require a cash-flow adequate to fund negotiations for what may be several years before any royalties are earned.

Patent-Management Organizations

The alternative preferred by many groups is the employment of a professional patent-management organization, providing access to skilled evaluative, legal, marketing, and patent-management expertise, with a minimum outlay of funds at the outset. Since income must be shared with the management organization, financial gain realized by the institution and the inventor will be reduced. From the institutional point of view, the potential reduction in income must be balanced against the estimated cost of maintaining an in-house patent-management staff. Unless your research staff produces a fairly constant flow of patentable results, establishing your own patent-management department can be fairly costly on a per-invention basis.

OMEC International, Inc. provides patent and license assistance, including invention disclosure and evaluation, strategic patent analysis and planning, patent filing and prosecution, prior art and trademark searches, patent alert service, and international licensing on inventions in biotechnology fields. Biotechnology has come to mean the use of microorganisms, plant or animal tissue cells, organelles, enzymes, or other constituents derived from cells to obtain a desirable transformation or beneficial product. OMEC's services are available to inventors in industry, academia, government, and private organizations.

Biotechnology Patent Digest, a biweekly publication of OMEC, is for scientists in industry, education, and government who wish to keep up-to-date on information contained in patents in biotechnology and related industrial fields. OMEC also provides a Patent Copy Service, offering printed copies of U.S. and foreign patents for a fee, based on speed of delivery. For information on OMEC and its publications, write to OMEC, Inc., 727 Fifteenth Street, NW, Washington, DC 20005.

Battelle Development Corporation (BDC) handles research, patent, and licensing functions for its parent institution, the Battelle Memorial Institute, a multinational, public-purpose organization with about 7,000 scientists, engineers, and supporting personnel. Battelle's research interests embrace the physical, life, and social/behavioral sciences. One of its major objectives is the benefit of mankind by the advancement and use of science through technological innovation and education. Since its incorporation in 1935, ten years after the founding of the Battelle Memorial Institute, BDC has sought to fulfill this objective by encouraging discoveries and inventions for economic use. BDC's expertise in patenting and licensing is available, on a negotiated basis, to institutions or individual inventors in industry and in educational and other nonprofit institutions. The address is 505 King Avenue, Columbus, OH 43201.

Research Corporation's logotype is a gear wheel that combines the academic lamp of learning with symbols representing the patent system and heavy industry. It may seem to be a strange pairing, but it is a perfect symbol for the corporation, founded in 1912. Its Technology Transfer Program has patented, developed, and marketed thousands of inventions and discoveries from academic laboratories, and schools of applied science, engineering and agriculture. Among those are vitamins A, B_1, and B_2, the laser concept, antibiotics, anticancer drugs, cortisone, the first heart-lung machine, the electrostatic precipitator, the cyclotron, and the Van de Graaff generator. Research Corporation enters into agreements with nonprofit institutions to provide evaluative, patenting, and licensing services on all discoveries submitted by the institution, on a royalty-sharing basis. For the year 1984, the program provided services to 292 institutions, mostly academic, and grossed nearly $10 million dollars in royalties. Approximately one-third of that income was distributed to institutions, and slightly more than that to inventors and other recipients. The program retained less than one-third. This pattern of distribution has been consistent for five years prior to 1984. A portion of Research Corporation's royalties is plowed back into research through their Cottrell Research Program and Cottrell College Science Program. The Research Corporation's main office is at 6840 East Broadway Boulevard, Tucson, AZ 85710–2815; and its Eastern Regional Office is at 44 South Bayles Avenue, Port Washington, NY 11050–3709.

University Patents, Inc. (UPI) enters into contractual arrangements with university laboratories under which UPI has the right of first refusal on essentially all research and development for a share of the royalties. In the more than dozen years UPI has been in existence, it has brought to the market an impressive list of products and the number is growing. UPI's service includes evaluative research, patenting procedures, and locating commercial licensees. Their operation was described in the September 1985 issue of *Nation's Business*, in an article entitled "Finding

Gold in the Ivory Tower." The address is University Patents, Inc., 1465 Post Road East, P.O. Box 901, Westport, CT 06881.

The Invention Submission Corporation assists individuals who may have a patentable "idea, invention, or new product," and they specify that it does not matter "whether the concept is brilliant, crazy, solves a problem, or provides a unique function." An initial screening is given at no cost, together with a pledge never to use any part of screened submissions without written permission of the inventor. The address is 1064 Clinton Avenue, Suite 130, Irvington, NJ 07111.

PATENT-INFORMATION SOURCES

Some reference sources of useful information about patents are:

Patent Law Fundamentals, 2d ed. New York: Clark Boardman Co., 1980. Updated biannually with replacement pages.

Massachusetts Institute of Technology and National Council of University Research Administrators. *Intellectual Property Series*. Washington, DC: NCURA, 1984.

Q & A About Patents, rev. ed. Washington, DC: U.S. Government Printing Office, May 1982.

Patents & Inventions: An Information Aid for Inventors. Washington, DC: U.S. Government Printing Office, June 1980.

Small Business Administration. *Introduction to Patents* (Management Aids Number 6.005). Washington, DC: U.S. Government Printing Office, 1963.

Chapter 10

LABORATORY SAFETY

Salus populi suprema lex. [The people's safety is the highest law.]
—Latin maxim

Your laboratory's research program is exceeded in importance by one thing only: safety—protecting your personnel and protecting the external environment from effluents, products, organisms, or hazardous substances that may be emitted as a result of your laboratory's research activity.

Not only is the protection of the health and safety of personnel and the environment a moral obligation, but because of an expanding array of federal, state, and local laws and regulations, it is a legal requirement as well. Industrial organizations, having learned their lessons well in the courtroom, often establish elaborate, well-staffed, and well-supported safety programs, organized primarily around production plants. Now similar safety procedures also apply to their research laboratories, and serious accidents caused by research activity are rare. Academic institutions have not been as diligent in the past, but as more university research involves radiation, carcinogens, and biotechnology, they, too, are beginning to develop sophisticated programs and enforcing safety policies more rigorously.

RESPONSIBILITY OF THE LABORATORY DIRECTOR

The responsibility for safety within a laboratory lies with the laboratory head and requires the active cooperation and participation of every member of the staff—from the newest, least-experienced employees to the most senior scientists.

In some ways, the administration of a safety program in a research laboratory is less difficult than in a production facility or even in a business office, because very early on, scientists learn how easy it is for a tiny misstep or variation in technique to create a disaster. There are, however, other factors that may hinder the smooth management of a safety program. One is that not everyone working in the laboratory is scientifically trained. The other is that researchers, often accustomed to working with hazardous substances or in dangerous environments and acutely sensitive to the care that must be taken at every step in the research procedures, may become disdainful of precautions against relatively mundane accidents (such as slipping on wet floors) that may result in serious injuries. Simply observe the attitude of scientists when forced to participate in routine building evacuation procedures for fire drills, bomb threats, and the like. Frequently, they strenuously resist and marshal convincing scientific arguments against leaving the lab at the sound of the alarm.

Laboratory directors must assign specific responsibilities for the safety program and be openly committed to its support. This commitment encompasses written support of the plan, organizational responsibility for it, and willingness to allocate personnel and financial resources to its implementation. The success of any safety program ultimately depends on the cooperation of the entire staff and the known interest and commitment of the laboratory head is the best way to assure full participation of all personnel. A program perceived as having only pro forma support from the top may be largely ignored by the rest of the staff.

The Safety Program

Essential elements of an adequate laboratory safety program are laboratory practices and techniques, safety equipment, building design, and trained personnel. The size of the laboratory and type of research it performs determine the management and size of your safety program staff. In a small laboratory mainly devoted to one type of research, the danger lies in assuming that, because the number of people and facilities involved are minimal, "somebody will take care of it."

But every laboratory, no matter the size, must institute an occupational safety program to establish and disseminate general guidelines; to see that each unit in the institution possesses the appropriate facilities—emergency exists, safety showers, fire extinguishers; to provide special clothing—lab coats, coveralls, safety glasses or goggles, gloves, or other protective items required; and to provide training and instruction in safety procedures and use of facilities, equipment, and protective devices.

The Occupational Safety and Health Administration (OSHA) of the United States Department of Labor is responsible for developing regulations to ensure the safety and health of people in the workplace. OSHA establishes and promulgates occupational safety and health standards, formulates and issues regulations, conducts investigations and inspections to determine the status of compliance with safety and health standards and regulations, and issues citations and proposes penalties for noncompliance. The laboratory director and the staff member responsible for general safety should be familiar with all OSHA regulations pertaining to the specific types of workers employed at your institution.

A waste-management system is essential for every laboratory, no matter what its size. Obviously, a system for a small college laboratory will be substantially different in detail from one designed for a large university or a large industrial complex, but there are certain fundamental elements that must always be present: (1) a laboratory director who is committed to the principles and practices of good waste management; (2) a well-developed and clearly articulated waste-management plan tailored to the operations of the institution; (3) assigned responsibility for the waste-management system; and (4) policies directed to compliance with the established practices and toward reducing the volume of wastes generated.

The plan should conform to the regulations of the Environmental Protection Agency (EPA) as well as those of state and local governments to control and abate pollution of the air and water by solid wastes, pesticides, radiation, and toxic substances. The plan should be prepared by those who are knowledgeable about the nature of the wastes and the means available for handling them and by those who will be responsible for the plan's implementation.

Safety committees representing personnel in each area who are likely to be exposed to various hazards must discuss problems, recommend safety measures, and facilitate communication and participation. Researchers who are intimately involved with hazardous substances must participate in creating policies, even though it may be necessary to have a full-time safety coordinator to select and oversee the installation of safety equipment, to ensure compliance with policies, coordinate emergency procedures, prepare reports, and monitor the effectiveness of the program.

SOLID WASTES

At one time nonhazardous solid wastes were useful as landfill. But as the need for landfill has declined, institutions are once again employing incinerators. These devices have been greatly improved with the addition of attachments designed to remove deleterious substances. Incinerator technology has also made great advances in converting waste materials into useful products. The American Society of Mechanical Engineers (ASME) holds a National Waste-Processing Conference and exhibit every two years and publishes the proceedings of these conferences. Publications derived from these meetings are available to nonmembers as well as members. Proceedings currently available are:

Waste Reduction = Conservation (Proceedings of the 1986 Conference). $125 (ASME members, $100).

Engineering: The Solution (Proceedings of the 1984 Conference). $125 (ASME members, $100).

Meeting the Challenge (Proceedings of the 1982 Conference). $100 (ASME members, $80).

Also available from ASME is *Thermodynamic Data for Waste Incineration*. Bk. no. H00141, 1979. 160 pages. $30 (ASME members, $15).

These publications may be ordered from ASME Marketing Department, 22 Law Drive, Box 2350, Fairfield, NJ 07007–2350.

Some wastes, such as those that emerge from medical-pathology laboratories, may need to be autoclaved before being incinerated and under no circumstances are they appropriate for landfill.

RADIATION SAFETY

The use of radiation sources in a research laboratory involves knowledge of, and compliance with, numerous federal, state, and local regulations. Moving nuclear products or disposing of nuclear wastes across state lines necessitate adherence to another level of regulation.

The U.S. Nuclear Regulatory Commission (USNRC) licenses and regulates the use of nuclear energy to protect public health and safety and the environment. It licenses persons and companies to build and operate nuclear reactors and to own and use nuclear materials and inspects the activities of persons and companies so licensed to ensure that no safety rules of the commission are violated. Some states have established a state radiation control agency licensed by the USNRC to handle monitoring and regulating radiation safety in those states.

The Environmental Protection Agency (EPA) develops national programs for air-pollution control, sets national standards for hazardous pollutants, and provides training in the field of pollution control. EPA gives technical assistance to states and agencies with radiation protection programs, and conducts a national surveillance and inspection program for measuring environmental radiation levels.

Moving nuclear wastes or nuclear products, particularly from one state to another, can be very complicated, involving the radiation control agencies of the states involved, the USNRC, the Department of Transportation, the EPA, and other federal or state agencies concerned with safety.

Research laboratories that use radiation from any source must have their radiation safety policies and practices established by an expert and the same person (or someone equally knowledgeable) must control access to the sources, supervise compliance with the guidelines, and regularly monitor personnel who may be exposed to radiation of any level.

Oak Ridge National Laboratory, one of the institutions where nuclear products were first developed, conducts a wide range of research involving radiation sources and provides USNRC with technical information to prepare environmental impact statements for energy facilities and for the supporting social and environmental assessments. No laboratory in the United States, or perhaps in the world, is more experienced and knowledgeable about the use of radioactive materials. The bibliography of references concerning radiation safety currently in use at that laboratory is shown on pages 152–153.

OAK RIDGE NATIONAL LABORATORY
RADIATION SAFETY BIBLIOGRAPHY*

TITLE 10, Code of Federal Regulations, Parts 0–199 Revised as of January 1, 1986.

Regulatory Guide 3.4, "Nuclear Criticality Safety in Operations with Fissionable Materials Outside Reactors," U.S. Nuclear Regulatory Commission (USNRC), Washington, D.C., January 1973.

Burchsted C. A. *Nuclear Air Cleaning Handbook*, ERDA 76–21 (1976).

Regulatory Guide 6.1, "Leak Testing Radioactive Brachytherapy Sources," USNRC, July 1974.

Regulatory Guide 6.2, "Integrity and Test Specifications for Selected Brachytherapy Sources," USNRC, July 1974.

Regulatory Guide 6.4, "Classification of Containment Properties of Sealed Radioactive Sources," USNRC, August 1980, revised edition.

Regulatory Guide 6.5, "General Safety Standards for Installations Using Nonmedical Sealed Gamma-Ray Sources," USNRC, June 1974.

Regulatory Guide 7.1, "Administrative Guide for Packaging and Transporting Radioactive Material," USNRC, June 1974.

Regulatory Guide 7.2, "Packaging and Transportation of Radioactively Contaminated Biological Materials," USNRC, June 1974.

Regulatory Guide 7.3, "Procedures for Picking Up and Receiving Packages of Radioactive Material," USNRC, May 1975.

Regulatory Guide 7.4, "Leakage Tests on Packages for Shipment of Radioactive Materials," USNRC, June 1975.

Regulatory Guide 8.1, "Radiation Symbol," USNRC, February 1973.

Regulatory Guide 8.2, "Administrative Practices in Radiation Monitoring," USNRC, February 1973.

Regulatory Guide 8.3, "Film Badge Performance Criteria," USNRC, February 1973.

Regulatory Guide 8.4, "Direct-Reading and Indirect-Reading Pocket Dosimeters," USNRC, February 1973.

Regulatory Guide 8.6, "Standard Test Procedure for Geiger-Muller Counters," USNRC, May 1973.

Regulatory Guide 8.7, "Occupational Radiation Exposure Records System," USNRC, May 1973.

Regulatory Guide 8.9, "Acceptable Concepts, Models, Equations, and Assumptions for a Bioassay Program," USNRC, September 1973.

Regulatory Guide 8.10, "Operating Philosophy for Maintaining Occupational Radiation Exposures as Low as Is Reasonably Achievable," USNRC, September 1975, revised edition.

Regulatory Guide 8.11, "Applications for Bioassay for Uranium," USNRC, June 1974.

Regulatory Guide 8.13, "Instruction Concerning Prenatal Radiation Exposure," USNRC, November 1975, revised edition.

Regulatory Guide 10.1, "Compilation of Reporting Requirements for Persons Subject to NRC Regulations," USNRC, October 1981, revised edition.

Laboratory Training Manual on the Use of Isotopes and Radiation in Soil-Plant Relations Research, Technical Reports Series no. 29, International Atomic Energy Agency (IAEA), Vienna, 1964.

Training in Radiological Protection: Curricula and Programming, Technical Reports Series no. 31, IAEA, 1964.

Design of Radiotracer Experiments in Marine Biological Systems, Technical Reports Series no. 167, IAEA, 1975.

Safe Handling of Radionuclides, Safety Series no. 1, IAEA, 1973.

Radiation Protection Procedures, Safety Series no. 38, IAEA, 1973.

Medical Supervision of Radiation Workers, Safety Series no. 25, IAEA, 1968.

Regulations for the Safe Transport of Radioactive Materials, Safety Series no. 6, IAEA, 1985 revised edition.

The Use of Film Badges for Personnel Monitoring, Safety Series no. 8, IAEA, 1962.

Basic Safety Standards for Radiation Protection, Safety Series no. 9, IAEA, 1982, revised edition.

The Management of Radioactive Wastes Produced by Isotope Users, Safety Series no. 12, IAEA, 1965.

The Basic Requirements for Personnel Monitoring, Safety Series no. 14, IAEA, 1980, revised edition.

Manual on Environmental Monitoring in Normal Operation, Safety Series no. 16, IAEA, 1966.

Techniques for Controlling Air Pollution from the Operation of Nuclear Facilities, Safety Series no. 17, IAEA, 1966.

The Management of Radioactive Wastes Produced by Radioisotope Users, Technical Addendum, Safety Series no. 19, IAEA, 1966.

Respirators and Protective Clothing, Safety Series no. 22, IAEA, 1967.

Basic Factors for the Treatment and Disposal of Radioactive Wastes, Safety Series no. 24, IAEA, 1967.

Manual on Safety Aspects of the Design and Equipment of Hot Laboratories, Safety Series no. 30, IAEA, 1981, revised edition.

Safe Use of Radioactive Tracers in Industrial Processes, Safety Series no. 40, IAEA, 1974.

Objectives and Design of Environmental Monitoring Programs for Radioactive Contaminants, Safety Series no. 41, IAEA, 1975.

Walton, G. N. *Glove Boxes and Shielded Cells for Handling Radioactive Materials*, Butterworths Scientific Publications, London, England, 1958.

*Provided by E. J. Frederick, Chemical Technology Division, Oak Ridge National Laboratory.

In addition to these sources, a recent publication (1986) of CRC Press contains detailed information on the management of radiation protection programs: *Handbook of Management of Radiation Protection Programs*, edited by Kenneth L. Miller and William A. Weidner, Milton S. Hershey Medical Center of the Pennsylvania State University. CRC Press, 2000 Corporate Boulevard, NW, Boca Raton, FL 33431. Prepublication price $96.

CHEMICAL SAFETY

Chemicals occur in almost limitless varieties, and under certain circumstances, virtually all chemicals can be hazardous. The professional researcher who performs chemical operations in the laboratory is usually acutely aware of the risks involved but yet may overlook some essential safety points. The condition of certain facilities—ventilation, safety hoods, incinerators, and showers, for example—may preclude the safe handling of some chemicals. What's more, scientists caught up in the excitement of their work may overlook the fact that other employees who come in contact with chemicals used in the laboratory are not adequately protected. Stockroom personnel, animal attendants, or those people making deliveries to the laboratory may all be exposed to chemical hazards unless appropriate precautions are taken.

OSHA regulations limiting the exposure of workers to chemicals are based on guidelines developed for about 500 substances by the American Conference of Governmental Industrial Hygienists. In general, levels are given that are not to be exceeded by the average exposure over an eight-hour day.

EPA is responsible for regulating chemicals in air, water, and land, including chemical disposal. As with radiation safety, there are state and local regulations that govern the disposal of chemicals and limit what can be discharged into sewer systems and the atmosphere, or placed in waste-disposal areas.

In the proliferation of information available from computerized databases, data on chemical substances have not been overlooked. MEDLARS (Medical Literature Analysis and Retrieval System) provides information on its database RTECS (Registry of Toxic Effects of Chemical Substances). It is an interactive version of an OSHA publication containing acute and chronic toxicity data for more than 57,000 potentially toxic chemicals. Records include toxicity data, chemical identifiers, exposure standards, and status under various federal regulations and programs. RTECS can be searched by chemical identifiers, type of effect, or other criteria. BRS (Bibliographic Reference Services) databases include HZDB (Hazardline), which provides access to safety and regulatory information on more than 3,000 hazardous substances. Information is collected from government agencies, court decisions, books, and journal articles. Data include information to help identify and handle hazardous substances.

Each department chairperson or laboratory supervisor must assume responsibility for the coordination of safety within his or her own area. Institutional safety coordinators can formulate guidelines, but it is impossible for a safety expert to be present when every chemical is removed from the shelf. Responsibility for safety during a specific experiment or operation lies with the individual in charge of the operation, including ensuring that all safety precautions are taken and, in case an accident does occur, seeing to it that it is reported immediately in the prescribed manner to appropriate authorities.

The Board on Chemical Sciences and Technology of the National Research Council Committee on Hazardous Substances in the Laboratory concerns itself with the problems of safe and practical handling and disposal of chemicals from laboratories. The committee has issued two reports, both available from the National Academy Press in Washington, DC 20418: *Prudent Practices for Handling Hazardous Chemicals in Laboratories* (1981), and *Prudent Practices for Disposal of Chemicals from Laboratories* (1983).

Robert M. Simon, senior staff officer of the Board on Chemical Sciences and Technology, advises that a third report is in press, to be released in 1987, on the handling and disposal of biohazards in laboratories.

BIOSAFETY

Microbiological and biomedical laboratories pose special safety problems for the laboratory director who must constantly conduct risk assessments of agents used and see to it that appropriate facilities are provided and good microbiological practices are observed so as to minimize risk.

Knowledge, techniques, and equipment to prevent most laboratory infections are now available, but the variety of indigenous and exotic infectious agents used in research is so great that no single code of practice, standards, or guidelines can serve every institution. And no single code can serve the same research institution over a period of time. The active microbiological or biomedical laboratory may be routinely dealing with new infectious agents or new strains of previously known agents, which, for every specific case, require a determination about the level of risk and method of dealing with it. Responsibility for making that determination sits squarely on the shoulders of the laboratory director. Risks associated with each agent, as well as with the activity to be conducted, must be considered and balanced against the potential value of the study or activity to be performed.

Classification of Etiologic Agents on the Basis of Hazard serves as a general reference source for some laboratory activities employing infectious agents. On the basis of this booklet and the concept of categorizing infectious agents and laboratory activities into four classes or levels, the National Institutes of Health and the Centers for Disease Control spent four years developing guidelines for biosafety in microbiological and biomedical laboratories. *Biosafety in Microbiological and Biomedical Laboratories* appeared in 1984 and includes a series of descriptions of standard and special microbiological practices, safety equipment, and facilities under four biosafety levels, as follows:

Biosafety Level 1: Practices, safety equipment, and facilities appropriate for undergraduate and secondary educational training and teaching laboratories and for other facilities in which work is done with defined and characterized strains of viable microorganisms not known to cause disease in healthy adult humans.

Biosafety Level 2: Practices, safety equipment, and facilities applicable to clinical, diagnostic, teaching, and other facilities in which work is done with the broad spectrum of indigenous moderate-risk agents present in the community and associated with human disease of varying severity.

Biosafety Level 3: Practices, safety equipment, and facilities applicable to clinical, diagnostic, teaching, research, or production facilities in which work is done with indigenous or exotic agents where the potential for infection by aerosols is real and the disease may have serious or lethal consequences.

Biosafety Level 4: Practices, safety equipment, and facilities applicable to work with dangerous and exotic agents which pose a high individual risk of life-threatening disease.

The practices and techniques, safety equipment, and facilities recommended for each level are summarized in the table on page 157.

Biosafety in Microbiological and Biomedical Laboratories also includes information regarding laboratory hazards and recommended precautions associated with pathogens that (1) are documented hazards to laboratory personnel; (2) pose a high potential risk to laboratory personnel; or (3) may produce diseases of grave consequences should infection occur.

Specific recommendations for work with an agent at a particular biosafety level are based on a risk assessment of the agent and of the planned activity. For example, diagnostic procedures with cultures of *Salmonella typhi* pose a primary infection hazard if the agent is accidentally ingested. Repetitive manipulations of small quantities of *S. typhi* would predictably not pose a hazard of infectious aerosol transmission. Consequently, the combination of standard and special microbiological practices, safety equipment, and facility features described for Biosafety Level 2 are recommended.

Hepatitis B virus (HBv) is among the most ubiquitous of human pathogens. Consequently, activities involving the use of HBv pose significant risks. Note that the incidence of infection in some categories of laboratory workers is seven times greater than that of the general population. Therefore, Biosafety Level 2 practices, containment equipment, and facilities are recommended for activities using known or potentially HBv infectious body fluids and tissues. For activities with high potential for droplet or aerosol production and for those involving production quantities or concentrations of infectious materials, precautions described for Biosafety Level 3 are recommended.

Biosafety in Microbiological and Biomedical Laboratories may be ordered from Superintendent of Documents, U.S. Government Printing Office, Washington DC 20402 (stock no. 01702300167-1, $4) or National Technical Information Service, U.S. Department of Commerce, 5285 Port

Summary of Recommended Biosafety Levels for Infectious Agents

Biosafety Level	Practices and Techniques	Safety Equipment	Facilities
1	Standard micro-biological practices.	None: primary containment provided by adherence to standard laboratory practices during open-bench operations.	Basic
2	Level 1 practices plus: laboratory coats, decontamination of all infectious wastes, limited access, protective gloves, and biohazard warning signs as indicated.	Partial containment equipment (Class I or II biological safety cabinet) used to conduct mechanical and manipulative procedures that have high aerosol potential which may increase the risk of exposure to personnel.	Basic
3	Level 2 practices plus: special laboratory clothing, controlled access.	Partial containment equipment used for all manipulations of infectious material.	Containment
4	Level 3 practices plus: entrance through changing room where street clothing is removed and laboratory clothing is put on; shower on exit; all wastes are decontaminated on exit from the facility.	Maximum containment equipment (Class III biological safety cabinet or partial containment equipment in combination with full-body, air-supplied, positive-pressure personnel suit) used for all procedures and activities.	Maximum containment

Source: Richardson, J.H., and Barkley, W.E., eds. *Biosafety in Microbiological and Biomedical Laboratories.* Centers for Disease Control and National Institutes of Health, USPHS, 1984.

Royal Road, Springfield, VA 22161 (stock no. PB84-206879, $6; microfiche $4.50).

Importation and interstate shipment of human pathogens and related materials are subject to the Public Health Service Foreign Quarantine Regulations (42 CFR Section 71.156) and to Interstate Shipment of Etiologic Agents (42 CFR Part 72). Information on the importation and interstate shipment of etiologic agents of human disease and other related materials may be obtained from Centers for Disease Control, Office of Biosafety, 1600 Clifton Road, NE, Atlanta, GA 30333.

Animal Pathogens

Laboratories engaged in working with animal pathogens should be aware that research involving livestock and poultry pathogens may require special laboratory design, operation, and containment features. A summary of the recommended levels for activities involving animals is given in the table on page 159.

Facilities for laboratory animals used for studies of infectious or non-infectious disease should be physically separate from other activities such as animal production and quarantine, clinical laboratories, and especially from facilities that provide patient care. Animal facilities should be designed and constructed to facilitate cleaning and housekeeping. A clean hall/dirty hall layout is very useful in reducing cross-contamination. Floor drains should be installed in animal facilities only on the basis of clearly defined needs. If floor drains are installed, the drain trap should always contain water.

Laboratory animal facilities, operational practices, and quality of animal care must meet the applicable standards contained in publications such as *Guide for the Care and Use of Laboratory Animals* (NIH publication no. 85–23, rev. 1985) and its appendix, "U.S. Government Principles for the Utilization and Care of Vertebrate Animals Used in Testing, Research, and Training," discussed in chapter 11 under the heading Research Animals, pages 171–179.

The importation, possession, use, or interstate shipment of some animal pathogens is prohibited or restricted by law or by U.S. Department of Agriculture regulations or administrative policies. For detailed information regarding restrictions on animal pathogens, write to Chief Staff Veterinarian, Organisms and Vectors, Veterinary Services, Animal and Plant Health Inspection Service, U.S. Department of Agriculture, Hyattsville, MD 20782.

Waste-management systems in microbiological and biomedical laboratories are especially critical and must be designed to deal with all kinds of hazardous substances (with the possible exception of radiation). In some laboratories radiation containment and disposal are also necessary.

A recent publication (1986) of importance in this connection has been issued by the American Society for Microbiology. It is *Laboratory Safety: Principles and Practices*, edited by Dieter H. M. Gröschel, John H. Richardson, Donald Vesley, Joseph R. Songer, Riley D. Housewright, and W. Emmett Barkley. It may be ordered from the American Society for Microbiology, Publication Sales, 1913 I Street, NW, Washington, DC 20006 for $51 ($38 for members).

As noted in the section on "Chemical Safety" above, the Board on Chemical Sciences and Technology of the National Research Council in

SUMMARY OF RECOMMENDED BIOSAFETY LEVELS FOR ACTIVITIES IN WHICH EXPERIMENTALLY OR NATURALLY INFECTED VERTEBRATE ANIMALS ARE USED

Biosafety Level	Practices and Techniques	Safety Equipment	Facilities
1	Standard animal care and management practices.	None	Basic
2	Laboratory coats, decontamination of all infectious wastes and of animal cages prior to washing, limited access; protective gloves and hazard warning signs as indicated.	Partial containment equipment and/or personal protective devices used for activities and manipulations of agents or infected animals that produce aerosols.	Basic
3	Level 2 practices plus: special laboratory clothing, controlled access.	Partial containment equipment and/or personal protective devices used for all activities and manipulations of agents or infected animals.	Containment
4	Level 3 practices plus: entrance through change room where street clothing is removed and laboratory clothing is put on, shower on exit, all wastes are decontaminated before removal from the facility.	Maximum containment equipment (Class III biological safety cabinet or partial containment equipment in combination with full-body, air-supplied, positive-pressure personnel suit) used for all procedures and activities.	Maximum containment

Source: Richardson, J.H. and Barkley, W.E., eds. *Biosafety in Microbiological and Biomedical Laboratories.* Centers for Disease Control and National Institutes of Health, USPHS, 1984.

1987 will issue *Prudent Practices for Handling and Disposal of Biohazards in Laboratories.* It will be available from the National Academy Press, Washington, DC 20418.

Recombinant DNA Technology

When the history of science in the twentieth century is written, it is hard to say which of three principal developments will be accorded the greatest significance. Will ours be known as the Computer Age? The Atomic Age? Or the Genetic Engineering Age?

On January 16, 1986, the world's first license was issued to market a living, genetically engineered organism to be used in agriculture, and less than four months later the license was suspended and sale of the organism was halted. The suspension resulted from charges by a concerned organization that the federal guidelines for allowing live, genetically altered microorganisms to be released into the environment had not been followed by the manufacturer. The organism involved was a virus, which had one gene snipped from its genetic code and was marketed as a vaccine to eradicate an outbreak of pseudorabies, a devastating herpes disease in swine spreading in the middle western United States. On July 23, 1986, the first genetically altered vaccine for humans was licensed by the Food and Drug Administration—a vaccine intended to protect against infection by the hepatitis B virus, a major cause of liver disease throughout the world.

No scientific breakthrough, with the possible exception of atomic fission, has created more fear, misunderstanding, and confusion mixed with hope among people than the discovery in 1973 of enzymes that permit splitting and recombining DNA in highly specific sites. The fear resulted in some frustrations for researchers but it had the salutary effect of impressing upon everyone in the scientific community, including academic, government, and private laboratories, that this work must be done under very carefully controlled conditions. Although a few researchers chafed under restrictions imposed upon them at the beginning (when public policy erred on the side of caution), a key problem for a time was impatience over delays in formulating the guidelines that were eventually published by NIH. Considering the complexity of the task, the NIH guidelines were developed with surprising expedience. Robert W. McKinney of the NIH Division of Safety has remarked:

> The development of guidelines for recombinant DNA research represents a milestone in the establishment of standard practices to be followed by scientists in dealing with a new technology. We understood immediately that standards had to be set and promptly went into action. This is a far cry from the way we dealt with nuclear energy technologies.

NIH guidelines are applicable to all recombinant DNA research within the United States or its territories conducted at, or sponsored by, an institution that receives any support for recombinant DNA research from the National Institutes of Health, including research performed by NIH directly. The guidelines are also applicable to work done abroad with NIH funds. If the host country has established rules for the conduct of recombinant DNA projects, a certificate of compliance with those rules may be submitted to NIH in lieu of compliance with the NIH guidelines, but NIH reserves the right to withhold funding if safety practices to be employed are considered inadequate.

Although NIH guidelines apply directly only to projects at institutions with NIH funding for DNA research, they have become the standard for many other laboratories, including those in the private sector. There are two main reasons for this. One is that some states, New York, for example, and at least one city, Cambridge, Massachusetts, have adopted these guidelines as legal requirement for conducting r-DNA activities in their jurisdictions. The other is that industry, now accustomed to the regulatory environment, generally tends to welcome standards by which its procedures may be judged—it eliminates differences from one area to another, and from one industry to another. Moreover, in our litigious society, it is important that standards be adhered to which can be cited as defense should legal questions ever arise concerning provisions for protection of people and the environment.

Jane K. Setlow of Brookhaven National Laboratory, who formerly chaired the NIH Recombinant DNA Advisory Committee (RAC), says, "By and large, the private sector has welcomed the NIH guidelines, and voluntarily adopted them, particularly in the pharmaceutical industry."

Institutional Biosafety Committee

The laboratory director sets up the administrative mechanisms to ensure the safe conduct of research. Surely guidelines such as those developed by the NIH for research involving recombinant DNA (r-DNA) molecules are invaluable, but no set of procedures can anticipate every possible situation. In the final analysis, research safety depends on each individual conducting the work.

The recombinant DNA guidelines for safe containment of agents used in r-DNA research categorizes hazardous substances in the four biosafety levels, as shown on page 157. Level 1 applies to the least hazardous agents and Level 4 to the most hazardous. The guidelines also stipulate that an Institutional Biosafety Committee (IBC) be established to ensure the safe conduct of research by instituting periodic reviews of safety provisions for proposed activities and establishing accountability for adherence to appropriate standards. For work with agents categorized as Levels 3 and 4, a Biological Safety Officer (BSO) must also be appointed and sit on the IBC. The IBC and the Principal Investigator (PI) determine the safeguards to be implemented for each project; and for those substances in the categories of Levels 3 and 4, the BSO must participate.

Each IBC must have no fewer than five members who collectively have the expertise to assess the safety of recombinant DNA research experiments and any potential risk to public health or the environment. At least two members must not be affiliated with the institution other than as members of the committee, but shall represent the interest of the surrounding community. Local public-health or environmental-protection officials, or those engaged in medical, occupational-health, or environmental work, are suggested.

The biosafety committee should include persons with expertise in recombinant DNA technology, biological safety, and physical containment. It should also include, or have available as consultants, persons knowledgeable in institutional commitments and policies, applicable law, standards of professional conduct and practice, community attitudes, and the environment, and at least one member from the laboratory technical staff. Standards established for the composition of the safety committees are such that they become competent bodies capable of monitoring safety for all microbiological research as well as that involving recombinant DNA molecules. The most recent *Guidelines for Research Involving Recombinant DNA Molecules* appears in the *Federal Register* for June 26, 1986 (vol. 51, no. 123).

The most direct and knowledgeable source of information at NIH covering these guidelines is the Office of Recombinant DNA Activities, National Institute of Allergy and Infectious Diseases, National Institutes of Health, Bethesda, MD 20205. Dr. William J. Gartland is the head of that office, and executive secretary of the r-DNA Advisory Committee.

For those seeking information covering the impact of biotechnology on hazardous waste treatment, especially for laboratories carrying on extensive microbiological research including recombinant DNA activities, it may be wise to purchase a copy of *Hazardous Waste Treatment: Impact of Biotechnology*, Report of OMEC Publishing Co., Washington, DC, 1986. It is available at $595.

Several federal departments, other than Health and Human Services, that have responsibilities for health functions are also concerned with the policies for conducting recombinant DNA research. The President's Office of Science and Technology Policy has established an interagency advisory committee for developing policy for federal agencies involved with the review of biotechnology research and products. It also provides a vehicle for communication and coordination between relevant departments, facilitates compliance with a uniform set of guidelines in the public and private sectors, and, where warranted, suggests administrative or legislative proposals.

The necessity for interagency coordination was demonstrated in the litigation concerning testing of a gene-altered virus to be used as a vaccine to prevent pseudorabies. The Department of Agriculture and the Environmental Protection Agency became involved when it was claimed that researchers had violated federal guidelines for the release of living gene-altered agents into the environment. (The suspended license was reinstated after appropriate coordination.)

Representatives of the following departments and groups are involved in working with this committee: Domestic Policy Council Working Group on Biotechnology, Biotechnology Science Coordinating Committee (BSCC), Department of Agriculture (USDA), Environmental Protection Agency (EPA), Food and Drug Administration (FDA), National Institutes of Health (NIH), National Science Foundation (NSF), and Occupational

Safety and Health Administration (OSHA). For further information about the work of this interagency committee, write to David T. Kingsbury, Assistant Director for Biological, Behavioral, and Social Sciences, National Science Foundation, 1800 G Street, NW, Washington, DC 20550.

*   *   *

Safety cannot be mandated and all the guidelines in the world are useless unless a commitment to providing maximum protection for every person who might be affected by an experiment is continuously present in the mind of every laboratory worker. While the principal investigator assumes the major responsibility for seeing to it that all work is performed with attention to standard safety practices, accidents can be caused by the least important member of your research team. Therefore, every member of the staff, no matter how peripherally involved in research, must be fully informed and instructed regarding all safety procedures.

Robert W. McKinney, who heads the Occupational Safety and Health Branch of the NIH Safety Department, is frequently invited to give lectures on biosafety. McKinney usually opens his talks by asking:

> In an organization that employs 500 people, what should be the size of the safety program staff?

In response, members of the audience give estimates ranging between two and ten. The answer, of course, is "Five hundred people." When accidents occur, it is seldom the fault of the plan; it is usually because someone failed to follow the appropriate procedures for carrying out a specific operation.

Chapter 11

HUMAN AND ANIMAL SUBJECTS

It is not what a lawyer tells me I may do; but what humanity, reason, and justice tell me I ought to do.

—Edmund Burke, Speech on conciliation with America, 1775

Scientific research involving human subjects has contributed to many advances—among them the development of machines to perform tasks swiftly and efficiently that once exhausted and sometimes killed men, women, and even children. It has led to the design of transportation vehicles easily operated by physically handicapped persons, freeing them from social isolation and rendering them more economically independent. Research has also led to better understanding of people as social beings functioning alone or in groups. Above all, advances in knowledge in various fields of biomedical science have led to new drugs, treatments, and surgical procedures that prevent and cure diseases, make patients more comfortable, prolong life, and ease pain in protracted illnesses.

The human race is deeply indebted to laboratory research animals; without them many biomedical and behavioral studies could not have been done. For example, when the National Institutes of Health suspended animal studies at Columbia University's health sciences division in January 1986 because of a failure to comply with guidelines that took effect as of December 31, 1985, the ban temporarily halted research on projects concerned with heart disease, cancer, AIDS, autoimmune disorders such as lupus, arthritis, infertility, organ transplant surgery, and birth defects such as Down's syndrome and cystic fibrosis.

HUMAN SUBJECTS

Research involving humans raises some troubling questions. Revelations about abuses of subjects in biomedical experiments during World War II led to drafting the Nuremberg Code of 1947—a set of standards for judging physicians and scientists who conducted experiments on prisoners in concentration camps. That code became the prototype for others, setting forth principles for the proper and responsible conduct of human medical research, the best known of which are the Helsinki Declaration of 1964 (revised in 1975), and the 1971 Guidelines (codified into Federal Regulations in 1974) issued by the United States Department of Health, Education, and Welfare. Codes for conduct of social and behavioral research have also been adopted, the best known being that of the American Psychological Association, published in 1973.

Protection

The Belmont Report, prepared by the National Commission for the Protection of Human Subjects of Biomedical and Behavioral Research, entitled "Ethical Principles and Guidelines for the Protection of Human Subjects of Research," was issued on April 18, 1979. The national commission was created by the enactment of the National Research Act on July 12, 1974 (Public Law 93-348). It was charged with identifying basic ethical principles that should underlie the conduct of biomedical and behavioral research involving human subjects and with developing guidelines to ensure that such research is conducted in accordance with those principles.

The Belmont Report lists three principles, or general prescriptive judgments, relevant to research involving human subjects. They are:

- respect
- beneficence
- justice

Respect for persons demands that subjects enter into the research voluntarily and with adequate information, i.e., that "informed consent" be given. While controversy prevails over the nature and possibility of informed consent, there is general agreement that the process must contain three elements: information, comprehension, and voluntariness.

The subject must be given sufficient information to provide a basis for the decision to participate in the research or to decline to do so. Such information generally includes the research procedure; the purposes, risks, and anticipated benefits; alternative procedures (where therapy is involved); and a statement offering the subject an opportunity to ask questions and to withdraw at any time from the research.

Special provisions must be made for subjects whose comprehension is severely limited (for example, infants and young children, mentally disabled persons, the terminally ill, and the comatose). A third party, such as a parent, guardian, or mate, must be chosen to act in the subject's best interest. That person should be able to observe the research as it proceeds and be able to withdraw the subject from the research, if such action appears in the subject's best interest.

The element of voluntariness becomes exceedingly delicate in some cases, and particularly when prisoners are involved. Prisoners should not be deprived of the opportunity to volunteer for research; however, under prison conditions it is possible for them to be subtly coerced or unduly influenced to engage in research activities for which they would not otherwise volunteer. This is one of those dilemmas that can be resolved only by balancing the competing values expressed in the respect-for-persons principle itself—whether to allow prisoners to "volunteer" or to "protect" them.

Beneficence, as a basic ethical research principle, implies that the proposed activity will do no harm, and that every effort will be made to maximize potential benefits of the research and to minimize any possible harmful effects. The Hippocratic maxim "do no harm" has long been a fundamental principle of medical ethics. Claude Bernard extended it to the realm of research, saying that one should not injure one person, regardless of the benefits that might come to others.

Unfortunately, learning what will, in fact, benefit others may require exposing some people to risk. The ethical approach is to decide when it is justifiable to seek certain benefits, despite risks, and when risks are so great that the benefits should be sacrificed. The application of this standard sometimes confronts researchers with very hard decisions. An example is research involving children that presents more than minimal risk without immediate prospect of direct benefit to the children involved. Some argue that such research is inadmissible, while others point out that this limit would rule out much research promising great benefit to children in the future. The application of the principle of beneficence may force some difficult choices.

Justice, as a general principle, means that people are treated equally, and in this sense it implies fairness in selecting those who receive the benefits of research and those who bear its burdens. During the nineteenth and early twentieth centuries, the burdens of serving as research subjects fell largely upon poor ward patients, while the benefits of improved medical care flowed primarily to private patients. In the United States in the 1940s, the infamous Tuskegee syphilis study used disadvantaged, rural black men to study the untreated course of a disease that is by no means confined to that population. The subjects were deprived of demonstrably effective treatment so as not to interrupt the project, long after such treatment became generally available.

Against this background, it can be seen how conceptions of justice are relevant to research involving human subjects. It is important to ascertain that some classes—welfare patients, particularly racial and ethnic minorities or persons confined to institutions—are not selected as subjects simply because of their easy availability or their manipulability, rather than for reasons directly related to the research. The Belmont Report urges that whenever research supported by public funds leads to the development of therapeutic devices and procedures, justice demands that these advantages do not go only to those who can afford them, and that subjects should not be selected from groups unlikely to be among the beneficiaries of the research findings.

Compliance with HHS Guidelines

The basic policy for the protection of human subjects currently in effect, established by the Secretary of the Department of Health and Human

Services (HHS) in accordance with the Public Health Service Act (as amended), is contained in the Code of Federal Regulations, title 45, Public Welfare, part 46, revised as of March 8, 1983 (45 CFR 46).

A "human subject" is defined by the Office for Protection from Research Risks, National Institutes of Health (OPRR-NIH) as a "living individual about whom an investigator (whether professional or student) conducting research obtains (1) data through intervention or interaction with the individual, or (2) identifiable private information." The regulations established by the OPRR-NIH extend to the use of human organs, tissues, and body fluids from individually identifiable human subjects as well as to graphic, written, or recorded information derived from individually identifiable human subjects. The use of autopsy materials is governed by applicable state and local law and is not directly regulated by 45 CFR 46.

There are some categories of research involving human subjects that are specifically exempt from these regulations. Exempt activities are those in which the only involvement of human subjects will be in one or more of the following five categories:

1. Research conducted in established or commonly accepted educational settings, involving normal educational practices, such as research on regular- and special-education instructional strategies, or research on the effectiveness of, or the comparison among, instructional techniques, curricula, or classroom-management methods.
2. Research involving the use of educational tests (cognitive, diagnostic, aptitude, achievement), if information taken from these sources is recorded in such a manner that subjects cannot be identified, directly or through identifiers linked to the subjects.
3. Research involving survey or interview procedures. However, if *all* of the following conditions exist, these activities are *not* exempt:
   a. Responses are recorded in such a manner that the human subjects can be identified, directly or through identifiers linked to the subjects.
   b. The subject's responses, if they become known outside the research, could reasonably place the subject at risk of criminal or civil liability or be damaging to the subject's financial standing or employability.
   c. The research deals with sensitive aspects of the subject's own behavior, such as illegal conduct, drug use, sexual behavior, or use of alcohol.
   (All research involving survey or interview procedures is exempt from IRB review, without exception, when the respondents are elected or appointed public officials or candidates for public office.)
4. Research involving the observation (including observation by participants) of public behavior, *except* in cases where *all* of the conditions listed above as item 3 exist; in those cases the exemption does not apply.

5. Research involving the collection or study of existing data, documents, records, pathological specimens, or diagnostic specimens, if these sources are publicly available or if the information is recorded by the investigator in such a manner that subjects cannot be identified, directly or through identifiers linked to the subjects.

Safeguarding the rights and welfare of individual subjects is, by policy of the Department of Health and Human Services (HHS), primarily the responsibility of the institution where the research is carried out.

Before HHS will expend any funds for research using human subjects, the applicant institution (any public or private entity or agency, including federal, state, and other agencies) must provide HHS with written assurance that it will comply with 45 CFR 46 regulations for research involving human subjects. This assurance must be executed by an individual authorized to act for the institution and to assume on its behalf the obligations imposed by the regulations. The written assurance to HHS includes the following:

1. The designation of one or more Institutional Review Boards (IRBs), whose function it is to review proposed research at convened meetings at which the majority of the members of the board are present. These review boards must have at least five members with varying backgrounds. They may not consist entirely of men or entirely of women or entirely of members of one profession. Each board shall include at least one member whose primary concerns are in nonscientific areas, such as law, ethics, or the clergy. There shall also be at least one member who is not otherwise affiliated with the institution and who is not part of the immediate family of a person who is affiliated with the institution.

2. A list of the IRB members, identified by name, earned degrees, representative capacity, indications of experience (such as board certification, licenses, etc.) sufficient to describe each member's chief anticipated contributions to IRB deliberations, and any employment or other relationship between each member and the institution (for example, full-time employee, part-time employee, member of governing panel or board, stockholder, paid or unpaid consultant). All changes shall be routinely reported.

3. A statement of principles governing the institution in discharge of its responsibilities for protecting the rights and welfare of human subjects.

4. Written procedures which the IRB will follow for conducting its initial and continuing review of research and for reporting its findings and actions to the investigator and to the institution.

A sample "Assurance of Compliance with the Department of Health and Human Services Regulations for the Protection of Human Subjects" can be requested from Office for Protection from Research Risks, PHS, NIH, Building 31, Room 4B09, Bethesda, MD 20892.

Institutional Review Board

Once an institution has an approved assurance from the Department of Health and Human Services (HHS), applications for research support from that institution to HHS must be reviewed and approved by the IRB and certification to that effect forwarded to HHS within 60 days after submission of application. General responsibilities for approval of a research project by the IRB are:

1. to determine that risks to subjects are minimized;
2. to ensure that risks to subjects are reasonable in relation to anticipated benefits or the importance of the knowledge that may reasonably be expected to result;
3. to see that the selection of subjects is equitable, taking into account the purposes of the research and the setting in which the research will be conducted;
4. to see that "informed consent" will be sought from each prospective subject or the subject's legally authorized representative;
5. to ensure that "informed consent" will be appropriately documented;
6. to make adequate provision for monitoring the data collected to ensure safety of subjects;
7. to provide for the protection of the privacy of subjects and the maintenance of confidentiality of data;
8. to establish additional safeguards where some, or all, of the subjects are likely to be vulnerable to coercion or undue influence, such as those with acute or severe physical or mental illness, or persons who are economically or educationally disadvantaged.

The IRB must assume additional duties when reviewing research involving fetuses, pregnant women, human in vitro fertilization, prisoners, and children. In all cases, it must be made clear to the subject, or the subject's authorized representative, that the subject has a right to withdraw from the experiment at any time, without penalty or loss of benefits to which he or she may otherwise be entitled.

Some categories of research involving no more than minimal risk to subjects may be reviewed by the IRB through *expedited review procedure*. Under such a procedure, the review may be carried out by the IRB chairperson alone, or by one or more experienced reviewers designated by the chairperson from among the other members; however, under this expedited review procedure, reviewers may exercise all of the authorities of the IRB, except that they *may not disapprove* the research. Before any research activity may be disapproved, it must be reviewed by the entire committee.

The most recent list of activities approved for expedited review includes recording of data from subjects 18 years of age or older using noninvasive procedures routinely employed in clinical practice—weighing, testing sensory acuity, electrocardiography, electroencephalography, and electroretinography. It does not include exposure to electromagnetic radiation outside the visible range (X rays, microwaves, etc.). It does include voice recordings made for research purposes, such as investigation of speech defects; research on individual or group behavior or characteristics of individuals, such as studies of perception, cognition, game theory, or test development, where the investigator does not manipulate the subject's behavior and the research will not involve stress to subjects.

Activities approved for review through the expedited procedures have been published in the *Federal Register,* and from time to time, as appropriate, the list is amended and republished in the *Federal Register.*

IRBs are required to document all activities, including attendance by members, votes on each action, and the basis for requiring changes in proposals. Minutes of meetings should be detailed, including votes for and against a proposed action and the number of abstentions. The IRB must retain copies of research proposals reviewed with the accompanying documents, such as the sample consent forms used, copies of all correspondence between the IRB and investigators, and progress reports of any injuries. These documents must be retained for three years after the completion of the research.

The regulations established by HHS for research projects supported by federal funds are appropriate guidelines for research projects involving human subjects whatever the source of funding. Federal, state, and local laws protecting humans used as subjects of research are consistent with the procedures required by HHS.

Those guidelines are available from the Office for Protection from Research Risks (OPRR), Public Health Service, National Institutes of Health, Building 31, Room 4B09, Bethesda, MD 20892. OPRR also has available copies of the report prepared by the National Commission for the Protection of Human Subjects of Biomedical and Behavioral Research, commonly called the Belmont Report. It was published April 18, 1979, and contains ethical principles and guidelines for the protection of human subjects of research upon which the current HHS regulations are based.

Investigators in charge of a research project have an ethical responsibility to see that the rights and welfare of human subjects of research in the laboratory under their administrative control are safeguarded, and to see that the institution's guidelines to ensure the protection of subjects are disseminated to everyone on the staff concerned with research.

Any scientist involved in research using human subjects should be aware of the Institute of Society, Ethics, and the Life Sciences, The Hastings Center, 623 Warburton Avenue, Hastings-on-Hudson, NY 10706. The institute publishes an annual bibliography, providing introductory references in the field of biomedical ethics, a sample of the technical literature, and discussion of specific issues in biomedical ethics. The center publishes a report six times a year with short articles on bioethics, and three times a year, a publication containing longer papers on current concerns of Hastings Center scholars. Articles and other publications of the center are valuable information and guidance for researchers.

Research Animals

Advocates of animal welfare continue to have a growing influence on the use of animals in research, and their activities have created some impediments for scientists. On the other hand, they have served society and the scientific community by supporting and sometimes prompting legislative action to set standards for breeding, selling, housing, transporting, feeding, and keeping records of laboratory animals, and for humane experimental procedures. Properly cared-for animals yield reliable research data and the cost of acquiring and maintaining healthy research animals is a wise investment.

The Animal Welfare Act (Public Law 89-544) was passed in 1966 and the guidelines issued in 1985 comprise the most recent of many subsequent amendments and implementing regulations. Their purpose is to set forth standards required when federal funds are used in animal research; the regulations implementing the standards are administered by the U.S. Department of Agriculture.

Proposed amendments to current rules and regulations implementing the Animal Welfare Act are published periodically in the *Federal Register* under the heading "Department of Agriculture Animal and Plant Health Inspection Service." Copies of the rules and regulations can be obtained from the Federal Veterinarian in Charge, Animal and Plant Health Inspection Service, in the capital city of most states, or by writing directly to the Department of Agriculture.

Guidelines have been established for government-supported research, which are appropriate for all research involving animals, whatever the source of the funding. They set high standards for breeding and care of animals and for the conduct of experiments. Adhering to these standards will ensure that your facility can meet federal standards, if public funding should ever be sought.

The guidelines are contained in NIH publication no. 85–23 (rev. 1985), *Guide for the Care and Use of Laboratory Animals.* Included as an appendix is "U.S. Government Principles for the Utilization and Care of Vertebrate Animals Used in Testing, Research, and Training." Copies of the *Guide,* together with the appendix, are available from the National Institutes of Health, Office for Protection from Research Risks, 9000 Rockville Pike, Building 31, Room 4B09, Bethesda, MD 20892.

Animal-Welfare Assurance

Institutions seeking government support of animal research projects must designate an appropriate high-level official who is ultimately responsible for animal care and use. The level suggested is a vice-president or chancellor with authority to commit the institution to meeting the requirements of federal policy.

The revised Public Health Service Policy on Humane Care and Use of Laboratory Animals by Awardee Institutions requires that applicants for federal support of activities involving animals must provide a written assurance of compliance with the Animal Welfare Act and the current guidelines regarding the use and care of laboratory animals. These assurances are submitted to the National Institutes of Health (NIH) Office for Protection from Research Risks (OPRR), 9000 Rockville Pike, Building 31, Room 4B09, Bethesda, MD 20892. The OPRR will provide your institution with necessary instructions and an example of an acceptable assurance. PHS funds will not be granted to any individual who is not affiliated with an organization that accepts responsibility for administration of the funds and that has filed the necessary assurance with the OPRR.

A key element in the assurance is the establishment of an Institutional Animal Care and Use Committee (IACUC). This is a generic name for the committee that must be established in accord with the PHS policy, but each institution may consistently use whatever name it chooses to assign to that committee, as long as the makeup and functions adhere strictly to the guidelines.

The National Science Foundation (NSF) requires grant applicants to submit a statement that the research has been reviewed and approved by the appropriate IACUC (or whatever the institutional committee is called), and that the grantee assures NSF that the PHS policy on human care and use of laboratory animals will be followed. However, NSF will accept applications from institutions not having a general assurance on file with OPRR, and will first review such applications for scientific merit. If a decision is made to support the proposal, the NSF will arrange for a special assurance to be negotiated.

Institutional animal-care committees must consist of not less than five members and include at least one doctor of veterinary medicine with training and experience in laboratory animal science and medicine, who has direct (or delegated) program responsibility for activities involving animals at the institution. One member must be a practicing scientist experienced in research involving animals; one must be a nonscientist, for example, a lawyer, an ethicist, or a member of the clergy; and there must be one member who is not affiliated with the institution in any way, other than as a member of the committee, and who is not a member of the immediate family of anyone affiliated with the institution. One committee member may fulfill more than one requirement—the individual who is not affiliated with the institution may also be one whose primary concerns are in a nonscientific area. However, the committee must consist of at least five members.

The committee is required to review the institution's program for humane care and use of animals at least once a year and to inspect all of the institution's animal facilities including satellite (off-site) facilities at least once annually.

Applications requesting federal support for activities involving laboratory animals must be submitted to the animal-welfare committee before submission to the funding agency, except for NSF as noted above. The committee must review the relevant sections of the application to determine that the proposed activities are in accordance with the policy of the PHS and that the provisions regarding the care and use of animals meet certain requirements.

All procedures with animals must avoid or minimize discomfort, distress, and pain to the animals, consistent with sound research design. If the procedures may cause more than momentary or slight pain or distress to the animals, they must be performed with appropriate sedation, analgesia, or anesthesia, unless the procedure without these ameliorating agents is justified in writing by the investigator for scientific reasons. Animals that would otherwise experience severe or chronic pain or distress that cannot be relieved will be painlessly sacrificed at the end of the procedure or, if appropriate, during the procedure.

The living conditions of animals must be appropriate for their species and contribute to their health and comfort. The housing, feeding, and nonmedical care of the animals will be directed by a veterinarian or other scientist trained and experienced in the proper care, handling, and use of the species being maintained or studied. Medical care for animals must be available and provided as necessary by a qualified veterinarian. Personnel conducting procedures on the species being maintained or studied will be appropriately qualified and trained in those procedures. Methods of euthanasia used will be consistent with the recommendations of the American Veterinary Medical Association (AVMA) Panel on Euthanasia unless a deviation is justified in writing by the investigator. (See 1986 Report of the AVMA Panel on Euthanasia, JAVMA, 188: 252–268, February 1986.)

Public Reaction to Animal Research Facilities. The animal-care committee of the institution must see to it that work in which animals are involved is conducted, not only in accordance with the policies established by the institution and with current legal requirements, but also *in line with local customs.* An NIH spokesperson, in commenting on the Columbia University suspension at the time it occurred, stated that this action was based on an inspection visit and also on complaints from the public.

Private organizations concerned about the welfare of animals can and do mount forceful compaigns in opposition to the use of animals in research. If people in the community become aware of animal experiments by hearing dogs bark, or by smelling a pig sty, or even by a sudden infestation of unfamiliar insects, they can express their opposition to the work of the institution in very embarrassing and troublesome ways.

The suspension of federal funding of a project at the University of Pennsylvania in 1985 followed four days of sit-ins by animal-rights activists at NIH headquarters in Bethesda, Maryland, and was partially based on evidence gathered in a 1984 raid on the university labs by the Animal Liberation Front. It was also the Animal Liberation Front that broke into the City of Hope Medical Center in Duarte, California, in December 1984, and took more than 100 animals, which they said were dying prematurely from neglect and poor treatment. The NIH investigation that followed led to the suspension of several million dollars worth of grants to City of Hope the following July. In some places public objections have made it necessary to abandon certain lines of research. This is expensive and demoralizing with a loss of potentially important data.

The Public Health Service continues to upgrade its requirements for the care and treatment of laboratory animals. And the federal government as well as industrial and academic laboratories are studying possible alternatives to animal testing, such as tissue-culture techniques and the use of computers.

The Congressional Office of Technology Assessment (OTA) recently issued a report entitled "Alternatives to Animal Use in Research, Testing and Education," based on a study conducted at the request of the Senate Committee on Labor and Human Resources.

The administrative head of the institution, as its public spokesperson and chief public-relations representative, can do a great deal to allay community fear by keeping the public well informed about the kinds of research being done and their ultimate value. This can be accomplished by, among other things, taking opportunities to appear on local radio and TV programs or by welcoming science writers and others to visit the laboratory when appropriate—of course, under circumstances that do not jeopardize the research or endanger your projects.

Personnel

The veterinarian in charge of your institution's animal-care facility oversees hiring and training of personnel, selection of animal suppliers, maintenance of animal-care facilities, and advises on experimental procedures, including administration of anesthetics and analgesics.

The staff veterinarian is also a key member of the animal-care committee and is particularly valuable in performing such functions as inspecting animal facilities and handling animal medical emergencies. Large institutions, with many research projects involving animals, may need several staff members trained in veterinary medicine. Among the deficiencies for which Columbia University's health science division was cited in 1986 was an inadequate number of veterinarians.

In conducting site visits to animal-care facilities during 1984, the NIH found that a number of institutions employed veterinarians on a part-time basis only. This arrangement may be adequate if the animal research projects are very limited and very small numbers of animals are involved, but the availability of a veterinarian on a part-time basis only puts severe constraints on the nature and scope of research that can be carried out satisfactorily and may hamper or delay the work of the institution's animal-care committee.

It is the responsibility of the animal-facility director to see that supporting staff and animal handlers are fully trained for their duties and to maintain a sound personnel health program. Transmission of disease between animals and humans is a serious concern for the personnel of an animal facility as well as for the animals. Even minor diseases that are not seriously hazardous to animals may have an effect upon the experiment and thus muddle the results. The availability of qualified, well-trained caretaking personnel is essential if research is to be conducted using laboratory animals.

Animal technicians are knowledgeable in the care and handling of animals, in basic principles of normal and abnormal life processes, and in routine laboratory and clinical procedures. These technicians function as assistants to veterinarians, biological researchers, and other scientists. For many years, the only way to acquire these technical skills was by on-the-job experience, but now there are places where good training is available, mostly in two-year programs with a high school diploma required for entry.

The American Veterinary Medical Association has been the accrediting agency since 1970 for educational programs in animal technology and will mail out, upon request, a complete list, by states, of accredited programs. Write to AMVA, 930 North Meacham Road, Schaumburg, IL 60196.

Graduate animal technicians have begun to form national organizations and set up continuing education programs, employment, and social services for the benefit of their members. These organizations are a useful resource for recruiting skilled animal-care personnel. AMVA also has a listing of state and local organizations of technicians.

Many states now require animal technicians to be registered or certified, and the specific provisions vary according to the laws of the various states. The American Association of Laboratory Animal Science (AALAS) provides examinations and registry for technicians who are eligible and employed in laboratory animal facilities. The AALAS is a clearing house for collection and exchange of information on all phases of the care and management of laboratory animals. It publishes several journals in the laboratory animal science field including *Laboratory Animal Science*, a bimonthly journal that prints papers presented at the annual sessions of the AALAS and other articles of interest to veterinarians and technicians. The AALAS Animal Technician Certification Board main-

tains a record of certified animal technicians, a valuable source of information about trained personnel. Write to AALAS at 210 North Hammes
Avenue, Suite 205, Joliet, IL 60435.

Animal Laboratory Facilities

No institution should embark upon a research activity involving animals
until the appropriate space and physical facilities are in place—not promised, or funded, or on the planning board, but ready for the animals to be
housed and cared for. At one time it was a fairly common practice to start
experiments involving animals by putting a few cages "here in the corner of the lab," or in the hallway, or the classroom. But today, because
of mandatory procedures, stories of such approaches are mostly apocryphal.

As a matter of fact, the standards for animal quarters in some places
are so high that they exceed those for human habitation in the same
institution. In his excellent article "Value and Ethics of Research on Animals," Neal E. Miller of Rockefeller University, commented, "If the in-
spector finds any cockroaches, or even flecked paint in my animal room, I
hear about it. But both exist in the Faculty Housing where I live. And
conditions in slum housing are vastly worse![1]

Appropriate facilities mean sufficient temperature-controlled space,
proper caging, healthy environmental conditions, maintenance of clean
surroundings throughout the building, good illumination, adequate feeding and watering equipment and facilities for cleaning and sterilizing it,
safe waste disposal appurtenances, and sanitary storage space for all food
and bedding supplies. Appropriate facilities also mean provisions for
separation of species and for quarantine and isolation facilities, labs for
pathological analysis of animal diseases, and set-aside space for surgical
and postsurgical care, treatment of disease, and euthanasia procedures.
They also include the designation of areas for administrative services and
for animal-care personnel to shower, change clothing, and relax during
break periods.

Since 1965, inspection and accreditation of laboratory animal-care
facilities has been a function of the American Association for Accreditation of Laboratory Animal Care (AAALAC). For accreditation purposes,
the AAALAC defines a laboratory animal-care facility as a commercial or
noncommercial facility maintaining, using, importing, or breeding laboratory animals for purposes of scientific research or investigation. The
AAALAC definition does not include those facilities for commercial production of animals not intended for use in research. A list of accredited
facilities and information on becoming an accredited institution can be
requested from the AAALAC, 208A North Cedar Road, New Lenox, IL
60451.

Acquisition of Animals

Animals used in research can be bred in the laboratory or purchased from commercial breeders. Breeding facilities are expensive to build and maintain, and, except in very special cases, it is usually better to purchase the animals from professional breeders. Facilities for transporting animals from the breeder to the laboratory must meet the same standards as facilities for housing the animals, and it is essential to have a quarantine area where animals can be held for an appropriate time before introducing them into an existing colony.

The design of experiments using animals is very precise as to the species used, age, weight, or genetic strain. Since most animal experiments are carried out over a period of time, animals with the required characteristics must be scheduled for delivery on exact dates. Therefore, in order for breeders to meet delivery dates with animals of the desired age, weight, and any other special characteristics, production schedules must be planned well in advance. And since the progress of experiments does not always go according to plan, it is essential that researchers and suppliers maintain close communication to coordinate deliveries with the progress of experiments.

Estimating the cost of a proposed project is easier when animals are purchased than it is if they are to be bred in the laboratory. Production is uncertain at best and can be affected by many factors. In writing proposals for grant support, it is a great help to have a firm cost estimate from a reputable breeder for the animals to be used.

The Animal Resources Program of the NIH Division of Research Resources helps meet the need of biomedical researchers for high-quality, disease-free primates and specialized animal research facilities. The program supports, via grants and contracts, primate research centers and their field stations, primate breeding and supply projects, development of animal models, animal diagnostic laboratories, and a variety of other animal research projects.

A research resources guide entitled *Animal Resources Directory* was completely revised for its sixth edition in 1985. It helps scientists find sources of assistance and collaboration involving animals in health research. The geographic index of the directory lists resources alphabetically by state and by title within each state and includes the names of cities in which they are located and the page on which each resource entry appears. The directory can be ordered from Public Health Service, National Institutes of Health, Building 31, Room 5B16, Bethesda, MD 20892.

It is not difficult to find the names of licensed and accredited breeders of research animals. The Veterinary Services Programs of the Animal and Plant Health Inspection Service, United States Department of Agriculture, publishes an annual listing by states of licensed dealers entitled "Animal Welfare: List of Licensed Dealers," which can be ordered from the Information Division, Animal and Plant Inspection Service, USDA, Washington, DC 20250.

AAALAC includes animal breeders in the institutions they accredit and in their listing of accredited facilities. It can be requested from AAALAC, 208A North Cedar Road, New Lenox, IL 60451.

Endangered Species

If research projects are designed to use animals not normally available from commercial breeders or exotic animals that must be imported, scientists must be aware of the Endangered Species Conservation Act of 1969, established to prevent import of endangered species of fish or wildlife into the U.S. and to prevent interstate shipment of reptiles, amphibians, and other wildlife taken contrary to state law.

Since the passage of the Lacey Law in 1900 (amended many times), researchers in the United States have been required to respect the wildlife laws of other countries as well. The consequences of failure to respect the wildlife laws of other countries was dramatically demonstrated in the summer of 1974 when Charles G. Sibley, director of Yale's Peabody Museum, was fined $3,000 by the United States Fish and Wildlife Service for importing for research purposes eggs of the peregrine falcon from Great Britain. The investigation had international ramifications: bird-egg collectors in England were fined, and a Danish museum official was fined and discharged.

Before designing or embarking upon a research project that proposes to use imported animals, researchers should consult the up-to-date *Endangered and Threatened Wildlife and Plants*, which includes both native and foreign species. The current list can be obtained by writing to the Office of Endangered Species, U.S. Department of Interior, Fish and Wildlife Service, Washington, DC 20240.

Alternatives to Use of Animals in Research

Public interest in animal welfare in conflict with public concern for meeting society's need for continued progress in biomedical and behavioral research led Senator Orrin Hatch, chairman of the Senate Committee on Labor and Human Resources, to request a study by the Congressional Office of Technology Assessment (OTA), entitled *Alternatives to Animal Use in Research, Testing, and Education*. The OTA assembled an advisory panel chaired by Arthur L. Caplan, of the Hastings Center, with members from animal-welfare groups, industrial testing laboratories, medical and veterinary schools, federal regulatory agencies, scientific societies, universities, and others—"representatives of all parties interested in laboratory animal use and its alternatives," according to John H. Gibbons, the director of OTA. The report, issued in February 1986, presents a range of options for congressional action using existing alternatives, developing new ones, disseminating research and testing information, restricting animal use, counting the numbers and kinds of animals used, establishing a uniform policy for animal use within federal agencies, and amending the Animal Welfare Act (which was amended by three laws passed in 1985).

According to the OTA report, most alternatives to current animal use in research fall into one of four categories:

1. *Continued, but modified, uses of animals.* Alleviation of pain and distress, substitution of cold-blooded for warm-blooded vertebrates, coordination among investigators, and use of experimental designs that provide reliable information with fewer animals than were used previously.
2. *Living systems.* Microorganisms, invertebrates, and the in vitro culture of organs, tissues, and cells.
3. *Nonliving systems.* Epidemiologic databases and chemical and physical systems that mimic biological functions.
4. *Computer programs.* Simulation of biological functions and interactions.

It is fairly certain that objection to use of animals in research will continue, and scientists are well advised to consider alternatives in planning future projects that in the past made extensive use of animals. The OTA report can be ordered from the Superintendent of Documents, U.S. Government Printing Office, Washington, DC 20402. The stock number is GPO–052–003–01012–7, and the price is $16.

An excellent source of general information on all aspects of using and caring for research animals is the Institute for Laboratory Animal Resources, National Research Council, 2101 Constitution Avenue, NW, Washington, DC 20418. Upon request, they will send a listing of available documents that can be ordered from the National Academy Press.

REFERENCES

1. Miller, N.E. Value and Ethics of Research on Animals. *Laboratory Primate Newsletter* 23(3):9, 1984.

Chapter 12

BUILDINGS AND EQUIPMENT

Size is not grandeur, and territory does not make a nation.
—Thomas Henry Huxley, "On University Education"

"**G**ood science is done in old buildings" used to be a common saying among scientists. And the belief that research workers should be required to construct their own apparatus continued until the early part of the twentieth century, even in some distinguished laboratories.

When Ernest (Lord) Rutherford became head of the Cavendish Laboratory in 1918, he expressed the opinion that the student who uses homemade apparatus, which always goes wrong, often learns more than the one who has the use of carefully adjusted instruments, which he is apt to trust and which he dares not take to pieces. When Rutherford's researchers complained (as many frequently did) of being starved for apparatus, he would reply, "Why, I could do research at the North Pole." And it is true that the apparatus that Rutherford, with the help of one laboratory assistant, used to induce the first man-made nuclear reaction in 1919 was exceedingly simple. As was the equipment used by Michael Faraday in the mid-nineteenth century, who made his remarkable electrical discoveries by constructing a voltaic pile with seven half-pence, seven discs of sheet zinc, and six pieces of paper moistened with salt water.

Alexander Hollaender also held to the belief that lack of equipment was no deterrent to doing good research. During the 1960s, when developing countries were imploring the United States to send them research reactors and other sophisticated equipment they considered essential to carry on education in the sciences, Hollaender used to say he could design enough experiments to keep a biophysics student busy for quite a long time using only a simple $5 UV lamp.

The question is not whether good science can be done in ancient buildings using simple handmade apparatuses. Of course it can. As Mao Zedong said in 1938, "Armament is an important factor in war, but not the decisive factor. . . . Man, not material, forms the decisive factor." But just as we no longer send warriors into battle armed with only their cunning and a lance, we cannot expect scientists to meet today's challenges of disease, renewable energy, world hunger, genetics, outer space, ecology, and so on, armed with only their intellect and a blackboard.

A White House panel, appointed to look into the health of colleges and universities, reported in May 1986 that among the major problems threatening the future of university research were obsolete equipment and aging buildings. The panel provided a set of recommendations aimed at strengthening university-based research for the next decade that included repeated calls for increased federal support. An article in the June 23, 1986 issue of *Business Week* noted:

> The universities, which conduct some 60% of the nation's basic research, are just beginning a painful process of rebuilding their outmoded facilities after years of neglect. About 240 of the nation's 290 engineering schools operate with substandard equipment, and educators estimate that it would take more than $30 billion to refurbish these facilities.[1]

Today's research tools are very expensive, requiring huge capital outlay. Educational institutions with reduced enrollments and falling income from other sources cannot provide and maintain the kind of facilities needed for modern scientific training and research. It has been estimated that getting *one* new chemist, physicist, or molecular biologist started in a laboratory typically costs between $50,000 and $75,000 in equipment alone. Research in high-energy physics requires high-energy accelerators that cost as much as $3 billion to build.

Mathematicians are the exception. S. M. Ulam, a mathematician who had a long association with John von Neumann and worked with physicists on the Manhattan Project, wrote: "Physicists (even theoretical physicists), biologists, and chemists need laboratories—but mathematicians can work without chalk or pencil and paper, and they can continue to think while walking, eating, and talking."[2]

Even though elegant and expensive instruments may not be required for every experiment or every kind of research, the availability of space and modern equipment has become a powerful recruiting advantage, as well as an incentive for top scientists to remain at an institution. The head of a research laboratory is, therefore, always under pressure to provide the research staff with state-of-the-art equipment, the most advanced cyclotrons, computers, ultracentrifuges, electron microscopes, synchrotrons, reactors, protein sequencers, DNA synthesizers, and other expensive instruments as soon as they are brought to market, or earlier, if possible.

In some places, laboratory space has become a measure of one's status, the way corner offices, carpeted floors, and windows assert a business executive's power. It is not only the amount of space that becomes important, but the design and configuration of the area that may have to be altered to meet the needs (or at least the desires) of a recently recruited researcher. It is good, therefore, to bear in mind in designing a new structure that a research laboratory is not a static piece of architecture, but may eventually require remodeling. Construction of new buildings, remodeling old ones, or additions to existing structures requires long-range planning. It is a mistake to design laboratory space with only the immediate need in mind; frequently, other needs may arise even before the original plan is completed.

LABORATORY BUILDING DESIGN

Scientists cannot design laboratory structures, and neither can architects. It takes both—preferably working together harmoniously with respect and tolerance for the special knowledge, talents, and needs of each other. Before an architect is consulted or selected, the governing body of the institution for which a structure is to be built must determine the fundamental principles upon which the design is to be based.

There are two dominant, and opposing, philosophies regarding scientific laboratories: one is that they should provide researchers with isolation from all outside interference—the "ivory tower" concept; the other is that they should provide opportunities to react with their surroundings or with other researchers to encourage imagination, providing for cross-pollination of ideas.

The Ivory Tower

One of the most spectacular examples of this concept was built on top of a mountain in Caracas, Venezuela, in the late 1950s—the Instituto Venezolano de Investigaciónes Científicas (IVIC). The staff was forced to park their vehicles at the foot of the mountain and be conveyed up to the laboratory on a special road—a tortuous road, tortuously built. One story—perhaps apocryphal—told about the construction of the roadway was that the laboratory director held the road workers at gunpoint when they threatened to strike, forcing them to finish their labor. When completed, the facility was beautiful, the equipment most modern, the staff competent—a veritable ivory tower. Although it may not have become the world's most notable laboratory, neither has it provided confirmation of the opinion of Sir Lawrence (W. L.) Bragg, who in 1915 shared a Nobel Prize in Physics with his father, Sir William Henry Bragg. Sir Lawrence compared the placing of pure scientists in well-equipped laboratories, divorced from teaching and application, with the assembly and isolation of poets in a fine house in beautiful country surroundings, where they were then told to get on with it. Under such conditions, pure scientists and poets would, as he put it, "not be visited by the muse." IVIC scientists have been sufficiently visited by the muse to bring honor and respect to the institute, certainly one of the finest research establishments in Latin America.

Interactive Plan

An example of the second philosophy is the New Cavendish Laboratory, designed in the late 1960s. Sir Brian Pippard, who chaired the building committee, laid down the guidelines for the structure in a document entitled *General Notes on the Proposed New Cavendish*. The guidelines emphasized the importance of compactness to promote communication between different groups. Although multistoried buildings commonly achieve this aim, there are serious disadvantages. For example, experimental work requiring heavy instruments and machines should be confined to the lower floors; such equipment demands reinforced flooring, and upper stories cannot be protected adequately from vibration. It was also recommended that, in addition to providing for large equipment on the ground-floor inside space, provision be made for the possibility of adding extensions or ells at almost any point, when additional heavy apparatuses were needed.

Pippard proposed a design of research blocks radiating out from a central core on the principle that wide blocks were preferable to long corridors, which were "psychologically unsatisfactory, deterring chance meetings and fruitful gossiping."

The general board of the university accepted Pippard's guidelines for the most part with some additions and revisions. A significant principle was also recommended: that some existing barriers between departments should be removed. In the opinion of the committee, the traditional divisions between disciplines had led to inflexibility, hampering economic and efficient deployment of limited resources.

Pippard's completed laboratory is a system of mostly windowless research rooms leading off from a central lofty, skylit hall, providing for chance meetings and idea exchange, but limiting opportunity for scientists to react to outside surroundings. At first, some staff members objected to the lack of windows, but soon realized they could easily walk into the open hallway and look outside. They also found that unbroken wall space was ideal for installing and attaching scientific instruments. Brian Pippard became the head of the Cavendish Laboratory in 1971. A few years after the lab moved to the new quarters, when asked about the expression "Good science is done in old buildings," he replied that the saying arose because most moves into new buildings are so badly planned that momentum is lost that can take years to reinstitute. For that reason, the move into the new laboratory was well planned, well organized, and the logistics handled efficiently. Pippard was not aware of any loss of momentum or falling off in the quality of research following the move to better quarters. "No basis for the myth!"

Combination or "Open" Plan

The Salk Institute in La Jolla, California, designed by the noted architect Louis I. Kahn in close consultation with the founding director, Jonas Salk, combines both philosophies—opportunities for isolation as well as interaction. The original concept of the institute envisioned a departure from traditional compartmentalization of the sciences and other disciplines, conventionally organized into departmental or other units. With such barriers eliminated, the whole of science and the humanities might become a unified force that could lead to a better understanding of the forces of nature and human life. The original fellows, who had been handpicked by Salk and persuaded to cast their lot with the new institution, were some of the world's outstanding scientists, interested not only in science but in the humanities and creative arts: the esthetic and philosophical side of man's life. Salk and Kahn designed a structure that they believed would foster the achievement of their goals.

Atop a bluff overlooking the Pacific Ocean, the two main buildings of the Salk Institute provide inspiring views from every laboratory. Each one can be entered directly from an outside walkway without passing through other areas. Walls along these walkways, however, are faced, at certain points, with blackboards instead of concrete, for the convenience of scientists who may encounter one another on their way to or from their labs. A large sunny courtyard joins the two buildings, where chairs and tables beside a constructed waterfall invite pauses for casual conversation. In addition to laboratory space, residential fellows are provided with separate offices where they can go to write or think or just to restore their souls. These "hideaways" do not have telephones unless a fellow requests one. Resident fellows are given budgets and yearly stipends with some additional perquisites. In-house research funds must also be supplemented with grants. Fellows have no teaching duties; their time is devoted entirely to research. Nonresident fellows—also among the world's best-known scientists—are invited to the institute for brief or extended stays.

When it was first occupied, the Salk Institute was some distance from any other buildings, although the University of California at San Diego (UCSD) and the Scripps Clinic and Research Foundation were nearby, allowing for frequent collaboration and shared seminars. Now UCSD has branched out to become a near neighbor. But as long as its western perimeter is the Pacific Ocean, the Salk will retain its sense of isolation.

The record of the institute speaks for itself. Whether the serenity of the setting coupled with the beauty and innovation of its structural design deserves any of the credit for its scientific achievements is surely debatable. What can safely be assumed is that the surroundings help enormously to attract some of the world's most noted scientists.

The judges who select winners of the "Laboratory of the Year Award," sponsored by the magazine Research and Development, show a clear preference for the open-plan design—one that encourages researchers to interact and also provides privacy. The twentieth annual award, given in 1986, went to Nabisco Brand, Inc.'s Robert M. Schaeberle Technology Center in East Hanover, New Jersey. It is a people-oriented laboratory, based on the recognition, according to the planners, that one could not tell researchers when or where inspiration might strike. Their goal was to build a structure that would further creativity, a place that would also be a pleasant environment in which to work. Among the laboratory's principal goals was to complete the construction as quickly as possible, at the lowest possible cost.

Nabisco asked the designers to strive for an energy-efficient building, yet staff comfort was to take precedence over energy savings. There was to be provision for future expansion and to accommodate unanticipated changes: spaces were to be versatile and adaptable to alteration. One requirement was to avoid having all personnel pass up and down a single corridor. The solution was the creation of a four-level, pinwheel-shaped structure, with three slender laboratory wings and an engineering office wing clustered around a central rotunda. The three-story building stands on gently rolling terrain and was designed to preserve tall stands of mature trees on the site. The main R&D level is housed on the first floor. The basement is used for manufacturing process testing and has a high bay area for tall apparatus. New food product research demonstration and testing are performed on the second floor. Engineering and marketing research functions are housed on the third floor. A circulation corridor leading to and from the rotunda extends the full length of each laboratory wing. There are also internal laboratory corridors. Wide, horizontal windows run along the circulation corridors and small, individual windows let daylight into each lab work area. Daylight is also brought in through clear-glass clerestories above all windows, and light shelves below. The light shelves also hold energy-efficient, fluorescent indirect-lighting fixtures.

The rotunda serves as a crossroads for the personnel circulation system, encouraging brief social encounters. Public areas, such as the technical-resources center, cafeteria, and training center, are clustered around this focal point; two freestanding elevator shafts provide vertical circulation for visitors. Openings between vertical service shafts provide entry to working labs. Lab benches are supplied with electricity, gas, compressed air, and vacuum lines. When needed, cryogenic and specialty gases are supplied in individual containers. Emergency showers and eyewash hoods are located wherever hazards might demand them.

The judges of the "Laboratory of the Year" also awarded high honors to the Monsanto Research Center in Chesterfield, Missouri, and a special-mention award was given to Schering-Plough Corporation's DNAX Research Institute of Molecular and Cellular Biology in Palo Alto, California.

The Monsanto site is also set on rolling hills with large stands of mature trees, creeks, and valleys, which were incorporated into the design to create a parklike setting. The research center includes a six-story plant-growth facility of 406,000 square feet and three main research buildings totaling 500,000 square feet. It was designed by architects and engineers working in close collaboration with the scientific staff. It contains 250 working labs, plus offices and conference rooms. Although lab modules and office layouts follow standards developed for other Monsanto buildings, innovations include data-handling centers on each floor, a hard-wire computer network throughout the building, a combination of variable-configuration labs, convertible fume hoods, and open-office plans. A central computer system also supervises operations and aids in management of energy and equipment. An open atrium and other public areas provide relief from the apparent sterility and sameness of the research areas.

The versatile features, such as demountable walls, convertible fume hoods, and adjustable furniture, for example, were designed with the expectation that the center will be an effective research facility for many years.

Configuring the Schering-Plough Institute was more of a remodeling or alteration problem than an architectural one; it was an interior design for an existing shell. Efficient use of space was achieved by arranging the labs around a central zone of shared activities and facilities. Large, open intersections of corridors encourage interaction of staff scientists, and the use of interstitial service runs facilitates maintenance and alteration of the basic laboratory fittings.

In summary, these features are thought to be desirable in laboratory building design:

1. *Privacy Coupled with Interaction.* Provision for the protection of the research staff from unwanted interference should be combined with ample opportunity to interact with the environment and other staff members.
2. *Heavy Equipment.* In multistoried structures, activities requiring heavy equipment should be located on the lower floors.
3. *Versatility.* Such features as demountable walls, movable furniture, and overhead cable trays for electrical and communications lines prolong the useful life of the laboratory.
4. *Illumination.* Illumination must be adequate for activities within each area.
5. *Comfort.* Efficient temperature and humidity control for personnel comfort plus the specific requirements for specialized research areas must be available.
6. *Space.* Wide aisles between lab benches are important. Space between and behind lab benches should be wide enough so that instruments can be powered from the rear and operators can pass between benches to service instruments.
7. *Service Lines.* Laboratory benches should be fitted with basic service lines, installed to facilitate maintenance without disruption of activity in the rest of the module. Interstitial service runs and overhead cable trays may be appropriate.
8. *Computer Facilities.* Computer systems can monitor experiments, enhance interdepartmental communication, supervise operations, and aid in energy and equipment management. Choose equipment appropriate for the size and nature of the institution, allowing for anticipated growth and expansion of computer needs.

9. *Shared Areas.* Shared facilities such as conference rooms, cafeteria, dispensary, stockroom, and general use equipment should be located in areas easily accessible to all.
10. *Safety.* Standard and statutory safety regulations must be taken into account; sprinkler systems, fume hoods, waste-disposal systems, personnel showers, and provisions for storage of food and drink outside those laboratories where contamination may occur should be installed as required.
11. *Handicapped Personnel.* Access routes for handicapped persons should be available.
12. *Construction Material.* Materials used in laboratory interiors must be compatible with the activities and the organisms used in each research module. Studies involving bacteria, viruses, yeast, tissue culture, or other biological substances may be affected by particular kinds of surfaces or by elements used in wall and floor finishes.
13. *Staff Participation.* Participation by the scientific staff in the design detail is essential.

These general rules are only basic guidelines for those charged with design responsibility. The amenities provided, or lacking, on the building site, geography and climate, size and nature of the scientific activity, and various other factors contribute to final decisions.

ROLE OF THE LABORATORY DIRECTOR IN CONSTRUCTION PROJECTS

It takes a particular kind of personality and talent to oversee construction of a new laboratory or to supervise a major alteration. Some superb research administrators quake at the thought of such an undertaking; however, there are those who not only willingly undertake the task but are veritable geniuses in persuading authorities of the need for new facilities and rallying the necessary financial and moral support. Administrators of this type have been described as having an "edifice complex." There is no situation in which their talents are so well and fully used as that which calls for rapid and steady construction or renovation. Their heads are filled with visions of more and yet more gleamingly new and efficient laboratories, with landscaped grounds, graded roadways, modern loading docks, improved ventilation systems, built-in state-of-the-art equipment, temperature-controlled chambers, centralized monitoring systems, and recreation and dining rooms. By the time a contract has been let for the current project, plans are on the drawing board for the next. They love to see structures rising and labs take shape. They are inspired by the sound of earthmovers, tractors, cranes, air hammers, riveting guns, the day-to-day excitement created by the progression of idea into blueprint into building. Impelled by this love of motion and growth, they are indefatigable in their quest for the necessary resources to keep the institution in a constant state of expansion. Their imagination and inventiveness know no bounds when it comes to convincing board members, funding organizations, or prospective donors of the potential achievement in research, if only the facilities were provided. A growing institution may be well served by such a leader, or at least a leader with a great deal of the "builder" in his constitution.

Robert R. Wilson, the first director of the Fermi National Accelerator Laboratory (Fermilab), in Batavia, Illinois, is known among his peers as a "builder" in the best sense of the word. In 1967, Wilson left Cornell, where the 12-giga electron volts (GeV) accelerator is named after him, to take over the National Accelerator, as the Fermilab was then called. He immediately pushed through completion of the 500-GeV accelerator on schedule at less than the funds appropriated for it and returned several million dollars to the federal government. William D. Metz, writing in *Science*, March 10, 1978, described Wilson's style of management as bold and aggressive:

> [He] became involved in the details of almost everything that went on in the laboratory. This included not only the physics of the accelerator building, but also the graphic design of the laboratory logo, the architectural design of several of the laboratory buildings, and the importation of a herd of buffalo to graze in the middle of the accelerator ring.[3]

Wilson's avocation was sculpture, and it was reported that Cornell University offered him a joint appointment in physics and architecture. Wilson resigned as director of the Fermilab in February 1978, when the federal budget for FY-1979 contained only $15 million for the project to double the peak energy of the facility from 500 to 1000-GeV. It was estimated that the facility would cost $30 million, and Wilson had threatened to resign unless the full amount were approved in FY-1979 so he could push through the project and meet the competition of the new European "Super CERN" accelerator.

Irrespective of the extent to which the head of the laboratory is involved in a construction project, there still must be a building committee, composed of scientists representing key disciplines or research areas. This approach ensures that the special needs of every group receive consideration. This is as true for major alterations as for new buildings and facilities. The building committee also serves as a check-and-balances system to prevent the domination of any one individual or any one research activity. The committee selects, or at least plays a major role in the selection of, engineers and architects who will design and carry out construction. The committee must have strong leadership. A scientist with engineering background or interest is often a good choice.

Only with the participation and active cooperation of the scientific staff can architects and engineers produce a structure that meets the needs of a particular research laboratory. Although building design alone will not make a great research institution, it can be a critical factor in the efficient use of research resources, assisting the research staff in bringing their ideas to fruition.

GENERAL-PURPOSE AND MULTIUSER EQUIPMENT

There are some essential laboratory instruments and equipment that are impractical or unnecessary to furnish to individual researchers—glassware washing and sterilizing machines, incinerators, radiation sources, literature-retrieval computer lines, animal-cage washing machines, copying machines, automobiles, trucks, among a long list of apparatus and services.

Because so many users are involved, decisions about model, manufacturer, size, number required, maintenance, location, and, for some facilities, scheduling of use, involve extensive deliberation, far lengthier than what may be needed to reach decisions about requests for instruments to be used by a single scientist or for one project.

Committees composed of users or representatives of users' groups must certainly be consulted, but equally important is expert advice to guide the final selection. Before experts are consulted, the committee may prepare a list of specifications:

1. *Purpose.* What is the purpose of the equipment? If the need is, for example, to transport heavy, unwieldy objects from one building to the next, or from one area to another, will a van or a light pickup truck do or will a larger vehicle be needed? If the problem is waste disposal, exactly what must be disposed of?
2. *Alternative Facilities.* Is there another existing facility that can be remodeled, altered, moved, or otherwise made to serve the purpose?
3. *Scale.* What volume of use is anticipated? How many trips? How much waste? How many users on the computer line, copying machine, radiation source? How many cages to be washed daily? How much expansion of need is anticipated?
4. *Sites.* What sites are available as possible locations? If none, what provisions must be made for housing and installing the equipment? If building alterations are required, have specifications for those been prepared, with details indicating changes and their affect on adjacent and other existing operations?
5. *Power and Service Lines.* Will additional power or service lines have to be installed?
6. *Maintenance.* Can the new apparatus be maintained by current laboratory staff or will a maintenance contract with the vendor or other outside firms be called for? If in-house maintenance is feasible, will the burden on the present staff be too great?

7. *Technical Skills.* Does the operation of the equipment require special skill? Is the skill presently available within the institution or will additional staff be necessary?
8. *Finances.* How much is the project going to cost? It may be too early for a detailed estimate, but approximate figures must indicate what range of expenditure is being contemplated.
9. *Consultants.* What kind of experts must be consulted?

The purchasing agent (see chapter 3, "The Administrative Staff/Purchasing") should surely attend some of the committee's sessions, and the plant engineer should also participate when certain details, such as load-bearing capacities, power lines, and resistance or susceptibility of certain spaces to noise and vibration, are to be discussed. It may also be necessary, when considering possible sites, to include representatives of groups not concerned with the new equipment but whose laboratories are near places being considered.

Once these and any other pertinent points have been explored, a statement can be drawn up describing the proposed equipment, the estimated volume of use, the disciplines to be served, the potential locations, any necessary structural alterations, and a rough estimate of cost. At this point, the purchasing agent or a committee member may be able to suggest appropriate vendors. It is, however, not time to call in a sales representative. Salesmen, like soothsayers, come in various guises, including among them friend, counselor, advisor, and consultant. They may well play all those roles, but they are unquestionably going to wind up recommending their own product no matter what the specified needs are. A representative of Company X is not likely to say, "We don't have our machine off the drawing board yet, but I understand Company Y has just brought out a fine one," or, "Ours is good, but Y's is better."

But salesmen, when they are technically reliable, can often be useful as your detective, with knowledge of many of the most up-to-date gadgets. It may be wise to maintain cordial relations with sales personnel from manufacturers of commonly used equipment, materials, and services. Frequently, sales representatives for scientific instrument companies are professionals, many with advanced degrees. As Alan Shaw of Biogen, SA, in Geneva, Switzerland has noted:

> They can and will help with all sorts of problems. Often, they can arrange novel purchasing and delivery schedules, set up standing orders, help track down missing shipments, negotiate attractive conditions for large orders and provide samples for testing—in addition to promoting their company's new products. The time you spend developing a working relationship with sales representatives is well worth it.[4]

When the items being considered represent a significant investment, such as incinerators, certain radiation sources, and large computers, it is worth making an additional investment to send the plant engineer and the member of the committee most concerned and knowledgeable about the need for the equipment to visit other institutions where similar work is being done and where they are currently using appurtenances you have in mind. But the particular item you are investigating should not be so newly installed that it is too early to evaluate effectiveness, maintenance problems, or dependability.

Preparation for visits to observe equipment in operation in other labs should be well planned. Those who plan to observe equipment in operation should go armed with questions culled from suggestions made by committee members and other interested parties. Inquiries must cover such things as operating efficiency, maintenance effort required, promptness of delivery and installation, disruption of services, personnel movement, or other normal research activity occasioned by the installation, deleterious effects on nearby activities, and overall costs, including maintenance.

The purchasing agent or plant engineer may know or be able to inquire about institutions where similar equipment is in use; if not, this may be the time to call on a consultant. Ideal consultants are those who are knowledgeable in the relevant field, unaffiliated with any commercial vendor, and unbiased in favor of a particular product. Professional societies may provide leads; government agencies that deal with energy and some of the regulatory agencies may be able to suggest specialists in particular disciplines. Nearby university faculties can be sources of experts. They must keep up with the newest techniques in order to properly train students, even though they may not have funds to purchase the most up-to-date item.

Once a consultant has been appointed and given the necessary background in order to study the matter and make a recommendation, a deadline should be set for submission of the final report and recommendation. The consultant's report must be presented to and discussed by the entire committee and a consensus reached. The final decision, however, is made by the head of the laboratory. Decisions that involve the commitment of significant sums of money are made after all the research, study, and recommendations are concluded. The laboratory head must then balance the good to be gained against the outlay of funds.

The newest, most-complicated, and costliest model often appears to be the best choice, but this is not always so. As Robert N. Ubell observed in *Physics Today*, August 1985, "When Dorothy, in the *Wizard of Oz*, finally arrives at the Emerald City, . . . the Wizard dazzles her with sparks, pushing buttons and throwing levers; but, unhappily, despite the Wizard's command of technology, he fails to send Dorothy back to her Aunt Em." Ubell concludes that scientific progress may be aided by powerful machines, but the greatest leaps are likely to continue to come from revolutions in thought.[5]

If capacity is a factor, and expansion of the need is anticipated, it is good economy to select a model with greater capacity than the immediate requirement. But if the need for increased capacity is not imminent, there is a possibility that a more efficient apparatus may be developed before the projected need becomes a reality. Modern technological development moves very rapidly and one must be always on guard against being bedazzled into the purchase of impressive pieces of equipment doomed to early obsolescence. On this point, Richard B. Setlow, associate director for life sciences at Brookhaven National Laboratory, commented, "I would not go so far as to say never buy state-of-the-art instruments, but the fact is that reliability is far more important than 'high class' when selecting scientific equipment."

A good example is solid waste disposal equipment. This field is in such a state of rapid development that careful research is needed to find the most efficient equipment to achieve the precise purpose for which it is intended. Until the 1960s, incinerators were widely used in industrial plants and research laboratories. Then along came the "Green Revolution," with better understanding about the dangers of polluting the environment and these were all but banished. Some types of nonhazardous solid wastes began to go into landfill. Now it has become apparent that there is a limit to our need for and our ability to absorb landfill. By 1985, it was estimated that some locations had only five or ten years of landfill use remaining. Concentrating predominately on the problems of municipal waste, mechanical engineers and designers have come up with a variety of improved incinerators, some with scrubbers to remove deleterious substances from the emissions. One proposed concept, aimed at "resource recovery," would burn wastes and produce steam and electricity. The American Society of Mechanical Engineers has a solid waste processing division that keeps abreast of the newest technology in this field and of theories that are being tested and disseminates information in *Mechanical Engineering* magazine.

Engineering and other professional society meetings often feature displays and demonstrations of the latest machinery and equipment of interest to the general membership. Those organizations concerned with animal research, for example, will invite exhibitors of cage washers, automatic watering systems, bottle-washing machines, and other equipment used in animal colonies. Many commercial companies, such as those promoting computers, have frequent exhibitions where their products can be inspected and observed in action. These demonstrations are promotional and must be viewed with certain reservations. It is particularly important to ask if a model being shown is presently on the market or if it is projected or being developed.

SCIENTIFIC INSTRUMENTS

Up to the twentieth century, scientific research was performed with very simple devices, usually constructed by the investigator with little or no assistance from others. The scientific knowledge accumulated with those crude instruments has lighted the way for today's highly sophisticated modern experimentation using elaborate machinery too complicated to be repaired, maintained, and, in some instances, operated by the researchers who make use of it.

Large-scale physics was said to have been born on the day in February 1932 when Cockcroft and Walton demonstrated that the apparatus they constructed with the help of T. E. Allibone was capable of disintegrating atoms by electrically accelerated particles. A few decades later, the term "big science," attributed to Alvin Weinberg, then head of the Oak Ridge National Laboratory, came into the vocabulary. It implied teams of researchers working with a variety of technically advanced instruments, powered by multiples of energy almost beyond imagination. The corollary then, "little science," or, more often (to avoid the derogatory connotation, one presumes), "small science" followed.

The nature of each research program determines the need for scientific instruments, but "big science" programs are, almost by definition, extremely expensive to equip. When the decision is made to launch a new project, or add a new line of research, a significant element of that decision deals with the provision for space, instruments, and services necessary to carry on the work. Approval of the necessary purchases is implied by the commitment to support the new activity. When the time comes to select the particular type, size, and model, with specific capabilities, that becomes the responsibility of the scientist in charge of the project.

In the normal course of research administration, requests for instruments are submitted annually when budget estimates are prepared, as part of the organization's capital budget. Depending upon the type of laboratory and its place in the hierarchy, the laboratory head may have little or no control over the annual amount allocated for capital equipment. Justifications submitted with budget requests may be sound and persuasive, but if the approving authorities send it back with the total slashed by fifty percent or more, the laboratory head is forced to make choices. That entails the very difficult task of setting priorities, deciding which needs are more urgent than others. It is not unusual for scientists to request a needed instrument several years in a row before gaining approval to buy it. Resourceful researchers do not allow such delays to prevent them from pursuing their investigations; they can often find ways to make up for the lack. They may use instruments located in another laboratory within the institution, even if it means the work has to be done at night or on weekends. Or they may find neighboring institutions where the apparatus exists and where spare usage time is available, or they may apply for a grant.

Research grants may include scientific instruments as a budget item, but that usually means standard items, equipment rental, or purchase of usage time, and not large expensive apparatus. Yet, it may also be possible to apply specifically for equipment or instrumentation grants.

NIH Shared Instrumentation Grants

In recognition of the long-standing need in the biomedical research community to cope with rapid technological advances and the rapid rate of obsolescence of existing instruments, the Division of Research Resources (DRR) of the National Institutes of Health (NIH) has set up a Shared Instrumentation Grant Program. Institutions that have received a grant through the Biomedical Research Grant Program are eligible to apply for funds to obtain instruments or systems that cost at least $100,000. Types of instrumentation supported include, but are not limited to, nuclear magnetic resonance systems, electron microscopes, mass spectrometers, protein sequencer/amino acid analyzers, and cell sorters. Support will not be provided for general-purpose or purely instructional equipment.

Applicants for the Shared Instrumentation Grants must be user groups of three or more investigators, all of whom must have Public Health Service peer-reviewed research support at the time of the award. The projects of the applicants must employ at least 75 percent of the total usage of the instrument; if there is not sufficient need at the grantees' institution for the remainder of the usage, access to the instrument can be made available to others under specified circumstances.

NSF Equipment Grants

Equipment grants for chemical, engineering, biological, and materials research and also for ocean sciences, earth sciences, mathematics, and computer research are offered by the National Science Foundation Equipment Program, as follows:

> Chemical Instrumentation: Division of Chemistry, NSF Brochure 86–17.
>
> Biological Instrumentation: Division of Molecular Biosciences; Division of Cellular Biosciences, NSF Brochure 83–19.
>
> Instrumentation for Materials Research: Division of Materials Research, NSF Brochure 85–66.
>
> Equipment Grants for Computer Research: Division of Computer Research, NSF Brochure 85–64.
>
> Engineering Research Equipment Grants: Engineering, NSF Brochure 84–51.
>
> College Science Instrumentation: Science and Engineering Education, NSF Brochure 86–23.
>
> Research in Undergraduate Institutions: Division of Research Initiation and Improvement, NSF Brochure 85–59.
>
> Science Computing Research Equipment for Mathematical Sciences: Division of Mathematical Sciences, NSF Brochure 85–44.

Earth Sciences Research Equipment: Division of Earth Sciences, NSF Brochure 85–26.

Facilitation Awards for Handicapped Scientists and Engineers: Division of Research Initiation and Improvement, NSF Brochure 84–62.

Ocean Sciences Instrumentation; Oceanographic Instrumentation; Shipboard Science Support Equipment: Division of Ocean Sciences, NSF Brochure 84–63.

Equipment Available at Other Institutions

Some types of research require such advanced instruments that no laboratory can provide all the necessary facilities using both internal and external funding sources. Recognizing this as early as four decades ago, the federal government began to construct and develop instruments for large-scale research, particularly in the nuclear sciences and related fields at laboratories supported by public funds. These facilities were created to anticipate and fulfill the obligation to scientists throughout the United States at first and, eventually, throughout the world. The national laboratories at Oak Ridge, Tennessee; Argonne, Illinois; Los Alamos, New Mexico; and Upton (Brookhaven), Long Island, have a constant flow of students and scientists, whose projects are relevant to the objectives of each laboratory, using the facilities for short or longer periods of time. For example, the High Flux Isotope Reactor (HFIR) at Oak Ridge; Brookhaven's National Synchrotron Light Source (NSLS); the Intense Pulsed Neutron Source (IPNS) at Argonne; and the National Stable Isotopes Resource (SIR) at Los Alamos, are all available to qualified outside users.

Proposals are also accepted for visiting scientists to use facilities built with public funds at places other than national laboratories, predominantly on college campuses. Examples include the 1-MeV electron microscope at the University of Colorado, Boulder; the Laser Microbeam Program (LAMP) at the University of California, Irvine; the National Nuclear Magnetic Resonance Facility for Biomolecular Research at the Massachusetts Institute of Technology, Cambridge; the 1.3 GeV Synchrotron Source at Lawrence Berkeley Laboratory in California; and the National Center for Biomedical Infrared Spectroscopy at Battelle's Columbus Laboratories in Ohio.

Technology Resources for Biomedical Research. The Division of Research Resources (DRR) of the National Institutes of Health publishes a directory, *Biomedical Research Technology Resources*, listing the facilities funded by DRR, all of which are available to qualified outside users.

Concentrating on the application of the physical sciences, mathematics, and engineering to biology and medicine, DRR supports the development of resources such as large-scale and minicomputer systems; biochemical and biophysical instruments (mass spectrometers, nuclear magnetic resonance spectrometers, electron spin resonance spectrometers); million-volt and scanning transmission electron microscopes; lasers; flow cytometers; vibrating probes; biomedical engineering technologies; and production of biochemical research materials. The complex computer systems are used primarily for statistical data reduction, mathematical analyses, biomedical modeling, and organized knowledge systems.

The directory is arranged by type of resource, for example, "Electron Spectroscopy Resources," and "Laser Resources," and has a geographic index. Each listing gives resource title, person in charge, services available, research emphasis or application, and the name and telephone of the user contact person. A copy of *Biomedical Research Technology Resources* (NIH pub. no. 85–1430) may be ordered from Research Resources Information Center, 1601 Research Boulevard, Rockville, MD 20850.

Technology Resources for Physics Research.   Physics and engineering research intersect with and are applied to every other scientific field. "Small" physics plays a central role in generating and elucidating fundamental physical and mathematical concepts. Condensed-matter physics, the major field of small physics, involves the study of materials by various types of spectroscopy and other techniques requiring fairly modest instruments, and is usually done in small groups. Atomic, molecular, and optical physics, and portions of nuclear physics, astrophysics, and plasma physics, and some others are effectively carried out in small groups. However, some aspects of small physics research require the use of large sophisticated facilities not normally available in research laboratories, such as synchrotron radiation and neutron sources, powerful electron microscopes, and large reactors. A brief description of twenty-three national facilities for research related to the physics of condensed matter that are available to qualified scientists from other laboratories is found in an article by George H. Vineyard and L. M. Falicov, which was published in *Review of Scientific Instrumentation*.[6] The facilities are grouped according to this outline:

   I. Neutron Sources
  II. Synchrotron Radiation Sources
 III. Facilities for Microanalysis, Microfabrication, and Surface Studies
  IV. Electron Microscopes
   V. Other Facilities

User-contact information is given for each facility listed.

Writing in the March 1985 issue of *Physics Today*, Daniel Kleppner of MIT reports:

> One telling indicator of the difference between small and big physics is the number of authors contributing to a scientific paper. The average

number of authors for a *Physical Review Letter* in condensed-matter physics is less than three; in particle physics it is more than 40.[7]

Research in high-energy and fusion physics, and much of that in nuclear physics, is performed at large facilities—the cost of one high-energy or fusion experiment is often in excess of $10 million.

High-energy physics research in the United States is funded primarily by the Department of Energy at such places as the national laboratories, including the Fermi National Accelerator Laboratory (Fermilab) and at universities. The Stanford Linear Accelerator (SLAC) with the storage ring SPEAR, at Palo Alto, and the Cornell High Energy Synchrotron Source (CHESS), Ithaca, New York, are publicly supported facilities, and therefore available to qualified outside users.

Resources of Brookhaven National Laboratory. The Brookhaven National Laboratory (BNL), operated by Associated Universities, Inc. (AUI), was formed in 1946 to seek new knowledge in the nuclear sciences and related fields through programs that require large-scale research tools, and to encourage the use of those large-scale facilities by scientists of university, industrial, and other laboratories.

Now, forty years later, BNL consists of more than 250 buildings and other structures and employs about 3,000 persons. Another 1,500 scientists and students from other institutions do research there for varying periods throughout the year.

Outside users who wish to retain title to any inventions resulting from work at BNL and to treat as proprietary all data generated during work at the facility, have the option to do so, provided they enter into a formal "Proprietary User's Agreement" with Brookhaven National Laboratory. In this case, the user pays the full-cost recovery to BNL for machine time and any related technical services which the laboratory provides. Industrial researchers are encouraged to make use of BNL's facilities and the Office of Research and Technology Applications has been set up to assist industrial researchers in making arrangements to do so. General information on the research facilities of BNL may be obtained from Brookhaven National Laboratory, Associated Universities, Inc., Upton, NY 11973.

Some of the facilities available for outside users at Brookhaven and the contact persons are:

Physics:

• *National Synchrotron Light Source (NSLS).* The world's brightest source of X-ray and UV radiation, for basic and applied studies in condensed matter, surface studies, photochemistry and photophysics, lithography, crystallography, small-angle scattering, and X-ray microscopy. Contact: Roger Klaffky, National Synchrotron Light Source, BNL, Upton, NY 11973 (516-282-4974).

- *High Flux Beam Reactor (HFBR)*. For the study of fundamental problems in solid state and nuclear physics and in structural biology and chemistry.

Contact: Roger Klaffky (see above).

- *Alternating Gradient Synchrotron (AGS)*. For basic research in elementary particle physics. Work began in 1986 to allow for a five-fold increase in AGS proton intensity and a twenty-fold increase of polarized proton intensity. It also allows for the acceleration of heavy ions up to and beyond gold and sets the stage for a Relativistic Heavy Ion Collider.

Contact: Derek Lowenstein, Accelerator Department, BNL,
      Upton, NY 11973 (516-282-4611).

Chemistry:

- *Positron Emission Transaxial Tomograph (PETT VI)*. Measures regional positron emitting isotope activity in the brains of human and animal subjects.
- *60-Inch Cyclotron*. Provides large quantities of carbon-11, fluorine-18, nitrogen-13, and oxygen-15, primarily for basic research. A four-particle variable energy machine, the cyclotron can accelerate protons, deuterons, helium-3, and helium-4 (alpha particles).

Contact: Alfred Wolff, Chemistry Department, BNL, Upton, NY 11973
      (516-282-4301).

Medical Research:

- *Neutron Activation Facility for Measuring Body Calcium*. Measures calcium, phosphorus, sodium, and chloride.
- *Neutron Activation Facility for Measuring Nitrogen*. Measures nitrogen, hydrogen, and fat by induced activation using prompt-gamma neutron activation analysis.
- *Whole Body Counter*. Measures total body content of gamma-ray emitting nuclides and induced radioactivity produced by neutron activation analysis.

The medical department also has an in vivo bone lead assessment facility and a mobile activation facility for measurements of cadmium or mercury.

Contact: Stanton Cohn, Medical Department, BNL, Upton, NY 11973
      (516-282-3591).

To find out about opportunities for industrial research at Brookhaven, contact William Marcuse, Office of Research and Technology Applications, Brookhaven National Laboratory, Upton, NY 11973 (516-282-2103).

Resources at the Fermilab. The Fermi National Accelerator Laboratory (Fermilab) consists of four miles of tunnel filled with sophisticated magnets and a 6,800-acre site where a wide array of engineering and physics disciplines are amalgamated. The technology-related programs at Fermilab include ion beams, superconductivity, cryogenics, laminar tooling, vacuum systems, high-power rf, computer simulation, beam optics, design with computers, computer controls, data processing, new computer architecture, electronic particle detectors, holography, solar energy, and neutron cancer therapy.

In 1980 the Fermilab Industrial Affiliates organization was established as a systematic attempt to bring the laboratory to the attention of industry and to find ways of effectively disseminating its scientific and technological developments. For information about the Affiliates or about technological developments at Fermilab, the contact persons are R. Carrigan (312-840-3200) and Henry Hinterberger (312-840-3395), Fermilab, Box 500, Batavia, IL 60510.

High-energy physics is a field in which scientists from all over the world work together in a common endeavor, and there is no better proof of this than the Fermilab. Of the 225 institutions represented in the laboratory's research program, about half of them are U.S. scientists and the other half come from other countries.

Resources in Other Countries.   American scientists also participate in physics research in foreign countries. The Institut Laue-Langevin, at Grenoble, France; the spallation neutron sources at Tsukuba, Japan and at the Rutherford-Appleton Laboratory in England; and the large CERN Center at Geneva have facilities that attract physicists from all the developed areas of the world including the United States. Martin Blume and David Moncton, in their article "Large Facilities for Condensed-Matter Science" (*Physics Today*, March 1985), list the major neutron and synchrotron radiation sources operating or planned throughout the world at the time of the writing.

* * *

Science, published by the American Association for the Advancement of Science, compiles an annual *Guide to Scientific Instruments,* which contains invaluable information about the latest developments in instrumentation. It is well organized, and is indexed by vendors. In the 1984–85 *Guide,* Earl J. Scherago stated in the editorial:

> In this age of increasing scientific specialization, one factor constantly reasserts itself: advances in instrumentation are perhaps the most important factors in opening new fields of science.

More information about the annual *Guide* can be obtained from *Science,* 1515 Massachusetts Avenue, NW, Washington, DC 20005.

USED EQUIPMENT

When an institution is awarded a government research grant or contract to perform work for which equipment must be obtained, it is a good idea to inquire whether the necessary pieces may be available through the "Excess Property" program of the particular agency for which the work is being done.

Federal agencies that award research grants and contracts to *profit-making* organizations usually retain title to any scientific equipment purchased with agency funds. When the work is completed, the property reverts to the granting agency, which may pass it along to another contractor or grantee.

When the agency has no need for the equipment, it is reported to the General Services Administration (GSA) as Excess Property. At regular intervals, GSA publishes a list, distributed to all federal agencies, of excess-property items. Any federal agency may request transfer to it of equipment listed. An agency may take title to property from this list and assign it to grantees or contractors to use it in federally sponsored research, usually at a cost to cover only the transportation to the research site. Educational institutions may acquire excess property under a variety of federal programs, the Higher Education Act, Vocational Education Act, and others.

After a specified time, items not taken over by another federal agency are declared to be Surplus Property and made available free-of-charge to eligible recipients. State and local governments, educational institutions, and most nonprofit organizations are eligible to receive surplus property. Most surplus property goes to state governments who then redistribute it to their state-supported institutions. States have what are known as State Agents for Surplus Property, who keep abreast of available used equipment and acquire whatever is useful for institutions in their states.

The National Science Foundation (NSF) retains title to a great deal of property because it offers grants to profit-making organizations. As a result, it can provide grantees with used instruments fairly often. The NSF also accumulates some pieces as excess property, which are reported to GSA and are published on the excess list. In 1986, grantees acquired over $18 million in excess scientific equipment through NSF's Grantee Excess Property Program.

The National Institutes of Health (NIH), the largest funder of biomedical research, provides grants and contracts mostly to nonprofit organizations, who are usually given title to equipment acquired in connection with the projects. Therefore, the NIH has very little used equipment returned by grantees and contractors. But NIH does have a large number of intramural research laboratories from which excess property may be derived. Apparatus declared excess from these intramural labs may be given to NIH grantees or contractors, or may be reported to GSA and appear on the excess list.

GSA is a central clearinghouse of the government for all excess and surplus property including, in addition to scientific instruments, office machines, automobiles and trucks, and other general-use equipment. Upcoming sales of government property are announced in the *Commerce Business Daily*, published by the Department of Commerce every weekday.

It stands to reason that the chances of getting a recent model or a sophisticated state-of-the-art scientific instrument through the surplus-property distribution chain are very slim, but there is a good chance of finding an older piece of equipment suitable, perhaps, for use in teaching. Much of the property donated by GSA goes to educational institutions through the responsible state agent. There is also a good chance of finding a discarded instrument or machine with good spare parts that may be expensive or even unobtainable in the marketplace. Instruments obtained as surplus property are frequently cannibalized by scientists, and the parts used in assembling unique apparatus that cannot be purchased.

PROPERTY MANAGEMENT

Responsibility for equipment management and inventory should be given to a specified member of the staff. Building equipment, such as heating and cooling machinery, utilities services, cleaning equipment, and motor vehicles, may be assigned to the plant engineering department. Servicing scientific instruments requires specially trained technicians who may be organizationally a part of the plant engineering staff, operate as an independent unit, or be administratively assigned to another department, such as the purchasing office.

Inventory

All laboratory equipment, both general-use and scientific, must be carefully inventoried and tagged with identifying information, including date of purchase, date of installation, maintenance instructions, and inventory number. Information on the location, age, and status of every piece of capital equipment should be available to the laboratory head on request. This implies, of course, that no instrument can be moved from one laboratory to another where a different supervisor is in charge without notifying the responsible office. And when laboratory or other supervisory positions change, transfer of equipment inventory responsibility should be a routine part of the transition.

Inventory information is useful as a guide to determine which equipment items must be replaced (based on the high cost of maintenance, owing to frequency of repairs, or based on depreciation records).

It is especially important that equipment provided by outside sources through research grants and contracts be carefully and accurately recorded. Equipment furnished to nonprofit organizations employing public funds usually becomes the property of the institution where the work is performed but problems can arise if the principal investigator on a grant moves to another institution before the research project has been completed.

When the principal investigator transfers from one nonprofit institution to another and takes the project along, the original institution has the option of deciding whether the equipment will be transferred. If the activity for which the equipment was purchased will not continue at the original institution (which holds title to the property), and the project will continue at the new location, the original institution usually agrees to a transfer. However, it must be kept in mind that certain instruments purchased on a grant for a particular project may be usable in numerous other activities within the laboratory. If your institution decides to retain the equipment, it has the authority to do so. If the institution wishes to continue the project, it also has the option (under certain grants) to name a substitute principal investigator to continue it. Your institution would then, of course, retain the required instruments and equipment.

If an entire project moves to another institution along with the principal investigator, grant funds may be used to defray the cost of moving the apparatus, depending upon the terms of the grant, the cost of the transfer, and the availability of funds remaining in the grant account.

Maintenance and Repair of Equipment

Laboratories with an extensive inventory of research instruments requiring attention to warranties, depreciation, maintenance, and security may find that record-keeping and maintenance scheduling is a full-time job. It is false economy to scrimp on maintaining expensive equipment. The life of instruments may be prolonged and their performance enhanced by establishing a regular schedule for services such as changing brushes on centrifuges, standardizing analytical balances, calibrating X-ray devices, and similar routine maintenance.

Custom-crafted or state-of-the-art equipment can usually be serviced most competently by the technicians trained at the facility where the apparatuses were designed and assembled. Since vendors have a stake in the efficient and reliable functioning of the items they distribute, they frequently offer maintenance contracts to their customers at the time such equipment is purchased. The plant engineer, purchasing agent, and technicians responsible for equipment maintenance in the laboratory should examine the technical requirements for servicing such instruments to determine whether it is more cost-effective to enter into a contract with the vendor or to rely on the internal staff. This assessment must include consideration of the cost of training staff technicians or adding new personnel and the level of performance to be anticipated from them, in comparison with the cost and established expertise of the vendor's representative.

REFERENCES

1. Smith, E.T. and Clark, E. Now R&D Is Corporate America's Answer to Japan, Inc. *Business Week* June 23, 1986. p. 136.
2. Ulam, S.M. *Adventures of a Mathematician.* New York: Charles Scribner's Sons, 1976. p. 52.
3. Metz, W.D. Fermilab Director Resigns: Cites Subminimal Funding. *Science* 199(4333): 1052–1053, 1978.
4. Shaw, A. Buying Wisely and Well. *Nature,* 300A:xvi, 1983.
5. Ubell, R.M. Editorial. *Physics Today* (Aug):BG7, 1985.
6. Vineyard, G.H. and Falicov, L.M. National Facilities for Research in the Physics of Condensed Matter. *Review of Scientific Instrumentation* 55(4): 20–30, 1984.
7. Kleppner, D. Research in Small Groups. *Physics Today* (3):79, 1985.

Chapter 13

FINANCIAL RESOURCES

There is only one proved method of assisting the advancement of science—that of picking men of genius, backing them heavily, and leaving them to direct themselves.

—James Bryant Conant, *Letter to The New York Times*

"The most important job of the boss of a laboratory is to keep the money flowing!" That is what Emanuel (Manny) R. Piore, former vice-president of International Business Machines Corporation used to say. As prosaic or cynical as that may sound, a good case can be made for its truth. No matter what other talents the head of a research institution may have, no matter how creative, intelligent, or respected in the scientific community he or she may be, if the organization does not remain financially viable, nothing else makes a difference. Just as a politician must first get elected before bringing about reforms or keeping a campaign promise, the laboratory head must achieve and maintain a financially stable organization before any research objectives can be realized.

The scientific administrator's fiscal responsibility can be stated quite simply: to acquire the necessary funds and to determine their use by allocating them among various scientific and nonscientific activities. Funding mechanisms at government, corporation, and university laboratories are strikingly different in many ways, but there are also numerous similarities. In the first place, all must justify to some higher authority the use of funds placed at their disposal. If the federal government does not perceive a need for certain activities of the national laboratories, or for a particular program being carried on at one of them, funding may cease. And if AT&T should decide that the Bell Laboratories are not contributing to the well-being and progress of the corporation, you can bet that Bell Labs would soon be out of business. Publicly supported colleges and universities are answerable to state legislatures, and private educational institutions and independent research laboratories (such as the Salk Institute) look to governing boards for approval and support.

INTERNAL FUNDS

Endowments, appropriations, tuition fees, and budgetary allocations from the parent organization or governing body might be referred to as *internal* funds, in order to differentiate them from support of sponsored research through grants and contracts, or *external* funds. Whatever the source of internal funds may be, it is usually necessary for the director of the institution to submit an annual budget request, to be prepared to defend that request, and to administer the approved funds in line with established policy.

An accountant (or an accounting department) is essential for budget preparation and budget monitoring. No laboratory director can become intimately involved in the details of budgeting, but neither can he or she function effectively without full and complete knowledge of the budget submission for every group or cost center in the institution. What's more, a system for keeping routinely informed of the rate of spending within each unit must be maintained. The detailed preparation of the budget submission is described in chapter 3, "Administrative Staff/Accounting," which indicates the steps that may take place before the director makes the final budget request.

Key elements in a budget request are amounts requested and justification. Budget preparation is, in fact, a program-review process. It is the point at which the laboratory head indicates which programs are to be emphasized by proposing to increase their allocations, which lines are to be downgraded or maintained at the same level, and what, if any, new programs are to be introduced. This is the point at which the director allocates the resources of the institution. Therefore each research project or activity must be thoroughly reviewed in order that requested amounts for the coming fiscal year accurately reflect program significance, taking into account, of course, the nature of the work and the size and kind of resources required.

Regular budget status reports, prepared by the accounting office, provide the laboratory director with a quick overview of the rate of spending in each cost center in relation to elapsed time. If a unit has spent fifty percent of its budget after only twenty-five percent of the fiscal year has gone by, there may be a problem. That is the primary use of the simple budget status report, which the director should get at least once each month; that is, it is an indicator of the general rate of spending by each cost center within each category of cost. Those accounts that seem to be on schedule need not be analyzed further, but those that signal an imbalance should be investigated. Overspending (in terms of elapsed time) does not always indicate trouble; excessive expenditures in the first part of the fiscal year may be required for a start-up after which costs may taper off. Or, underspending in the first months may reflect anticipated expenditures in the latter part of the year.

A detailed, midyear budget review is generally a good idea; it is usually not too late then to make adjustments by transferring funds between cost centers or within cost centers, to institute austerity, or to request additional funding.

EXTERNAL FUNDS

When endowments, appropriations, and other internal funding sources decline, one of the most effective ways to augment the budget is through grants and contracts from public or private funding agencies. It is also the best means of funding programs and activities for which no internal funds can be made available.

External funding has many beguiling attractions, accompanied by some pitfalls. In fulfilling the primary responsibility for keeping the institution financially viable, the leader can never compromise its scientific integrity. This is not easy, even in the best of times. In difficult times, when funds are scarce, it is exceedingly tempting to accept outside projects that may otherwise have been rejected. When this occurs, "sponsored research" comes to have a pejorative, second-class connotation. When a laboratory undertakes a research project primarily for the money it brings in, not only is the quality of research compromised, but something happens to the administrative standards as well. The management of outside funds is likely to be somewhat casual, even careless at times. This has been implied for years by the use of the terms "soft money" when referring to grant funds, and "hard money" when speaking of the institution's regular budgetary income.

The willingness to lower the criteria for research projects in order to increase income provides the perfect climate for the rise of the entrepreneurial scientist—the "good grantsman." An apt description of such a scientist is contained in a statement made by one, "You tell me where the money is, and I'll write a proposal to get it!" When the institutional head is hard-pressed for funds, these go-getters may look like lifesavers but, in fact, they may threaten the life of the institution. If they are permitted to seek outside research support that promises income but brings little else to the institution, pedestrian projects soon absorb a disproportionate amount of the laboratory's resources, such as space, equipment, support personnel, and administrative time.

The mechanism of outside funding, properly used, however, has numerous advantages other than financial increase. It is an indicator of the significance of the work of the laboratory to outsiders; it brings that work to the attention of experts who serve as reviewers for grant applications; and it is one means of evaluating the work of individual researchers. There is perhaps no better evidence for scientific excellence than having one's work recognized by scientific peers who participate in selecting projects for funding. The key to making proper use of the sponsored-research mechanism lies in formulating and adhering to a research policy that does not tolerate a double standard. If a project does not meet the standards of the institution or contribute toward its overall goals, it should be unacceptable, no matter how much money is involved.

Control of sponsored programs must remain in the hands of the laboratory director, who is responsible for establishing and maintaining a quality research program. Certain aspects may be delegated to trusted and competent subordinates, but the final decision as to whether a particular project is acceptable to the institution comes from the top. In fact, most granting agencies insist that the institution where the work is to be done act as the "grantee" and administer the funds. The grant may be awarded on the basis of the knowledge, skill, and scientific reputation of the principal investigator, but the funds are usually paid to the institution, which is then held responsible for their management according to principles stipulated in the award document. Therefore, scientists who undertake projects and their institutions are partners in externally funded projects, from the moment preparation of an application begins until the final fiscal closeout after the completion of the work.

It is therefore essential, as in all partnerships, that each party understands the point of view of the other and that they present a united front to the external party, the granting agency. In order to present a united front, the institutional head and the principal investigator must be clear about their joint obligations to fulfill the work sponsored by the outside source (while simultaneously adhering to the highest standards of the laboratory).

If researchers understand the goals and ideals of the laboratory, they can more easily discriminate between activities that will advance those objectives and those that will sidetrack or deter them. The laboratory's policy should define those research areas that are of major importance and therefore ought to receive special emphasis and encouragement. Further, it should designate the less-than-major or peripheral areas of institutional interest; these may relate in some way to the major concerns, or they may have special geographic or other current significance. The policy should also deal with methods for maintaining a balance in the research program in order to avoid allocation of a disproportionate share of the talent and facilities to one area and thereby threaten major objectives.

In formulating and implementing your institution's research policy, it is indispensable that the senior research staff be consulted. A committee composed of those individuals can assist not only in formulating policy but in establishing guidelines and procedures to ensure that all research proposals emanating from the laboratory are consistent with that policy.

Many universities (and some research organizations) have established offices for handling all grant and contract matters. The primary function of these offices may be to assist in increasing the flow of outside funds into the institution, but in the most enlightened situations, the grants and contracts administrator is sufficiently familiar with institutional policies to review proposals on two major points: (1) to determine their appropriateness for the institution from a scientific point of view— their consistency with the tenets of the research policy; and (2) to determine that the obligations that will devolve upon the institution as a result of the award are reasonable and can be met.

An experienced grants and contracts administrator may be able to mediate inconsistencies and resolve differences that arise between the principal investigator's view and the institutional policy, but if questions arise that cannot be settled between them, the head of the institution is always the final arbiter. If decisions are exceedingly difficult, as they may be in some gray areas, the matter may be referred to a research advisory committee made up of senior scientists working in or knowledgeable about the relevant field.

But no matter how the review process is handled, once the proposal is finally signed by the head of the institution or the authorized official, that signature indicates to the funding agency that the proposed project is of a nature, quality, and size that can be carried on in that laboratory; that the necessary services and facilities will be provided; and that the institution understands all commitments to which it is subject as a result of the proposal and is willing to fulfill those commitments.

The following checklist, separated into Research Matters and Operational Matters, may be helpful as a guide for the institution in reviewing proposals and for the principal investigator in preparing applications, which must have the endorsement of the institution before they are submitted to a funding agency.

External Funding Checklist: Research Matters

1. *Main or Peripheral Area.* Does the subject matter fall into the main area or one of the main areas of research in which the laboratory is currently active? If not, does it fall within a peripheral area that overlaps with a major one?
2. *Personnel Qualifications.* Are the qualifications of the personnel to be involved sufficient to carry it out successfully, and have those qualifications been fully documented?
3. *Available Personnel.* Will the necessary personnel be available?
   a. *Obligations*
      Are those indicated as research participants too overburdened with other obligations—academic, research, and so forth—to fulfill the role described for them? If they are to be released from other duties in order to participate in the project, has this been agreed upon by all concerned?
   b. *Outside Consultants*
      Is heavy reliance on outside consultants implied? If so, what evidence is there that those individuals are available and willing to participate?
   c. *New Personnel*
      Must new research personnel be added to the staff? If so, is there reasonable assurance that they can be recruited and present at the time their contribution will be required? Are there hiring restric-

tions—personnel "freezes"—that might make it difficult to bring in new people?

    d. *Nonscientific Personnel*

      Can the other personnel needs be met—technicians, administrative staff, and so forth? Are they available within the institution or must they be recruited from outside? If recruitment is anticipated, has this been fully explored by the principal investigator and a plan made for handling it?

4. *Size.* Is the size of the project appropriate for the institution? Will it dislocate a disproportionate share of the talent and facilities of the institution? This is especially important if the subject is in an area of peripheral interest.

5. *Interference.* Will it interfere with other ongoing research? Bacteria, viruses, or radioactive substances may affect other types of research within the same laboratory, building, or even the same complex of buildings.

6. *Quality Standards.* Does the proposed activity meet the standards of quality with which the institution is identified and which it wishes to maintain?

7. *Community Reaction.* Are there likely to be adverse reactions from the community due to the nature of the research or some aspect of its performance? Criticism may arise from factors such as fear of accidents that would affect the community, chemical odors, noise, use of publicly owned facilities, or failure to recognize and deal with the requirements for the protection of animal and human subjects. If so, what measures will be taken for dealing with these adverse reactions?

8. *Facilities Burden.* Will the additional burden placed upon the facilities, the personnel, or fiscal resources of the institution be justified on the basis of the potential value of the research to be done or the results obtained?

9. *Animals.* Will animals be used in the project? If so, has a source for obtaining them been identified, and has the laboratory committee responsible for reviewing proposals involving animals approved of the provisions made to ensure compliance with the governmental regulations regarding animal welfare?

10. *Human Subjects.* Will human subjects be used? If so, has the proposal been reviewed and approved by the laboratory committee responsible for compliance with governmental regulations concerning participation of humans in research?

External Funding Checklist: Operational Matters

1. *Space.* Is the necessary space available?

    a. *Requirements*

      Are the estimated space requirements reasonable in terms of the needs of the project and the personnel to be involved?

b. *Internal Space*

If space within the institution is to be used, has it been ascertained that it can be assigned to the project when it is needed? There may be as much as a year or two between application for funds and start-up of work under a grant or contract; therefore, *projected* availability, not *current*, must be assured.

c. *Outside Space*

If outside space must be acquired, has preliminary exploration indicated that it can be obtained, and have reasonable cost estimates been included in the budget?

2. *Instrumentation.* Will specialized instruments be needed? If so, and if they are already within the institution, will they be accessible to the project staff when the time comes for their use? If they must be purchased, have provisions been made in the budget based on a reasonable estimate of the cost at the time they will be ordered? Cost of equipment changes rapidly; current price quotations must be obtained (last year's catalog won't do) and allowances made for increases in price due to inflation between date of application and date equipment is to be ordered.

3. *Specialized Facilities.* What specialized facilities are required: computers, cyclotrons, quarantine and holding quarters for animals, parking space or overnight accommodations for visiting subjects, clinical testing facilities, and so forth? If any of these are necessary for the proposed activity, preliminary arrangements must be made and reasonable assurance given that they will be available to the project director when they are required according to the projected time schedule of the work.

4. *Other Institutions' Facilities.* Will the facilities of any other institution be required? If so, the willingness of that institution to participate in that project and the agreed-upon basis of that participation should be documented in the proposal.

5. *Administrative Services.* What additional workload will be placed upon the institution's administrative service units, such as the personnel, purchasing, and accounting offices or the maintenance and construction departments?

6. *Other Facilities.* Will the project increase the use of other facilities, such as the library; the editorial, graphics, or visual-aids departments; the cafeteria; the passenger or service elevators?

7. *Salaries.* Do the salary scales reflected in the budget conform to the scales of the institution for all personnel and meet the standards of the funding agency?

8. *Fringe Benefits.* Have provisions been made for fringe, vacation, and sick benefits for personnel to be employed on the project?

9. *Affirmative Action.* Have all requirements been met regarding compliance with civil-rights and affirmative-action laws?

10. *Patents.* Have appropriate provisions regarding patent agreements been included?

11. *Matching Grants.* Must the institution share in the cost of the project on a matching basis? If so, and the match is in cash, what provisions have been made for obtaining the necessary funds? If the match is a "soft match," that is, participation in the costs by providing in-kind services or facilities, has this been properly coordinated within the institution by consultation and agreement with all those involved?

12. *Overhead Costs.* Does the budget reflect the institution's current negotiated rate? Or, in lieu of a negotiated rate, do the calculations accurately reflect the allowable overhead costs?

An adverse response to any of the questions may not necessarily make the proposal unacceptable or inappropriate for your institution, but the proposed research that meets the criteria indicated by favorable compliance on all these points promises to provide benefits to the institution and to the research community that outweigh the obligations it will place upon the organization where the work is carried out.

The matters raised by these questions are those about which the granting agency will want information; therefore, the application that responds fully to those points will stand a better chance of success. Using such a checklist as a guide, the likelihood of misunderstandings and friction between the administration and the researcher will be greatly reduced, and will go a long way toward preventing the institution from being overtaxed or embarrassed with an inappropriate project.

Reimbursement for Overhead Costs

Government granting agencies and many foundations recognize that there are two types of costs involved in research projects. *Direct* costs are those readily identifiable with the project, such as salary, fringe benefits, supplies, equipment, travel, publication costs, and so forth. *Indirect* (or overhead) costs are those incurred by the institution in providing support services that are shared by other activities, such as general administrative expenses, including the time faculty and department heads spend administering federal research projects, utilities and various service lines, radiation facilities, animal quarters, libraries, janitorial services, and so forth. Institutions must be reimbursed fairly for that portion of general services necessary to carry out the provisions of the grant which cannot be identified and funded as a separate element in the direct-cost portion of the budget. The federal government refers to this reimbursement as "indirect cost" and has a procedure for negotiating a *fixed rate* for educational and other nonprofit institutions that is applicable to most federally funded projects. Responsibility for negotiating this rate has been assigned to the Health and Human Services department (HHS). The HHS comptroller in each of the ten federal regional offices is the authorized negotiator for institutions within each region. Most federal agencies honor the fixed rate set by HHS.

In June 1986, the Office of Management and Budget (OMB) revised that portion of Circular A-21 (the document that determines how institutions are reimbursed for indirect costs associated with research grants) covering reimbursement for the time department heads and faculty spend overseeing federal research projects. That reimbursement is now limited to three percent of direct research costs. This is a windfall to those places where reimbursement for salaries of supervising faculty and administrators has been less than three percent but is not welcome news to those where the percentage has been higher. However, "effort reports," in support of administrative contributions to government-funded projects have been eliminated, and that is good news for everybody.

Foundation board members are often insensitive to the concept of overhead cost reimbursement, mainly because they are unaware of the burden that some funded activities can place on the facilities of the institution. They have a tendency to see a grant as a gift which enlarges the income of the organization. But many foundation program officials do understand the concept and can be very helpful by advising on the presentation of a budget in a way that is acceptable to foundation directors and still includes an amount to cover administrative and other indirect costs.

Corporations have an appreciation of overhead costs and usually use the contract mechanism to sponsor research in laboratories outside their own facilities. They expect to pay the actual cost of the project, but they are quite meticulous, as a rule, about having every item clearly spelled out. They are, in general, not receptive to including a blanket amount for general institutional overhead but have no objection to specific items, such as administrative services, that are normally subsumed under overhead or indirect costs.

For the past twenty years, the federal government has included in some of its research grants a cost-sharing requirement; a certain percentage, often very small, of the direct cost of a project had to be borne by the grantee institution. That requirement was deleted for Public Health Service research grants beginning in February 1986. Existing institutional cost-sharing agreements remain in effect through February 3, 1987. This change applies to research grants and does not affect such cost-sharing requirements as matching funds for construction grants.

Sources of External Funds

In the halcyon days following World War II and, in fact, up through the 1960s, the acquisition of research support through grants and contracts was surprisingly easy, especially in comparison with today. However, it is still essential for most research institutions to find outside sponsors in order to maintain an active and vital research program and most certainly in order to train the next generation of research scientists.

The major sources of outside funds for research are the federal government, corporations, and foundations. But the government is, by far, the most lucrative source; it gives more for research than the others combined. The Public Health Service (PHS), through the National Institutes of Health, funds biomedical research, and through its other agencies, funds research related to the biomedical sciences. The 1987 PHS budget request was nearly $5 billion. The National Science Foundation (NSF), with a budget approaching $2 billion (FY 1987 budget request, $1,685.7 million), was originally established in 1950 to support basic science, but since 1980 has expanded to include engineering and applied science as well. Prior to that time, pressure had been building for research seeking solutions to practical problems, culminating in a congressional proposal at one point that some money be taken away from the NSF to create a new Technology Foundation.

In 1980, John B. Slaughter was appointed as NSF's first director with training as an engineer. Since September 1984 Erich Bloch, formerly an IBM engineer, has directed the NSF. The 1987 budget request reflected continued support of basic research but showed specific emphasis in biotechnology, computational science and engineering, global geosciences, and broadening participation in science and engineering research.

NSF currently offers support in numerous programs including mathematical and physical sciences; engineering; astronomical, atmospheric, earth, and ocean sciences; advanced scientific computing; as well as programs in biological, behavioral, and social sciences. NSF also makes equipment grants for chemical, biological, earth sciences, computer, and materials research. Deadlines for submission of applications to the NSF are given in the *NSF Bulletin* published monthly (except July and August) by the National Science Foundation, 1800 G Street, NW, Washington, DC 20550.

Corporations support a great deal of research and training at educational institutions, mainly using the contract mechanism. Much of their support is likely to go for research in technological and applied fields, but there has been mounting corporate support for projects in the social and behavioral sciences, too.

A few foundations support scientific research in medical and biomedical fields, but it is not significant. Foundation giving in 1985 for health and hospitals was 14.1% of their total giving, approximately $11.25 billion; the remaining $68.59 billion was divided between religion (47.2%), education (13.8%), social service (10.7%), arts and humanities (6.4%), civic and public programs (2.8%), and 4.9% for all others. The largest portion of moneys going to health and hospitals was for training health professionals and not for research.

Every laboratory should have in the library (or some other central location) at least a few basic references on grants and contracts, including, if your budget permits, periodical publications and newsletters that focus on funds available in specific research fields. It is better to have a small reference shelf that is kept current than to have many volumes that are out of date. Grants and contracts information gets old very fast. Most of the basic information sources are updated annually, or at least every two years.

The following basic references and periodicals publications are suggested for a core grants and contracts library:

Government. *Catalog of Federal Domestic Assistance.* A comprehensive listing and description of federal programs and activities which provides assistance or benefits to the American public. U.S. Government Printing Office, Washington, DC 20402. The 1986 subscription price: $30 domestic; $37.50 foreign. Subscription consists of a basic manual and supplementary material for an indeterminate period. In looseleaf form punched for three-ring binder. Designed to assist in identifying the types of federal domestic assistance available, describing eligibility requirements for the particular assistance being sought, and providing guidance on how to apply for specific types of assistance.

Federal Register. Published daily, Monday through Friday, by the Office of the Federal Register, National Archives and Records Administration, Washington, DC 20408. Provides a uniform system for making available to the public regulations and legal notices issued by federal agencies, including those affecting existing grant programs and proposed new ones. U.S. Government Printing Office, Washington, DC 20402. Subscription: $300/year; $150/six months.

The United States Government Manual. The official handbook of the federal government, prepared by the Presidential Documents and Legislative Division of the Office of the Federal Register. Information on the agencies of the legislative, judicial, and executive branches of the government, including principal officials and a brief history of the agency. The last section, "Sources of Information," for each agency provides information on contracts and grants and other areas of public interest. U.S. Government Printing Office, Washington, DC 20402. The 1986–87 edition price: $19.

Commerce Business Daily. Published daily Monday through Friday by the Commerce Department, charged by law with publicizing all proposed procurements of $10,000 or more by civil and military agencies, all federal contract awards of $25,000 or more (for the benefit of potential subcontractors), and all foreign government procurements. These include requests for bids and proposals, procurements reserved for small business, prime contract awards, and federal contractors seeking subcontract assistance. Upcoming sales of government property are also announced. U.S. Government Printing Office, Washington, DC 20402. Subscription: $243/year (priority); $173/year (nonpriority).

NIH Guide for Grants and Contracts. Published weekly (as a general rule) to announce opportunities, requirements, and changes in grants and contracts activities; to announce scientific initiatives; and to provide policy and administrative information to individuals who need to be informed about the National Institutes of Health. Notices of the availability of Requests for Proposals (RFPs) are published. Eligible recipients receive this publication free. For information, write to Guide Distribution Center, National Institutes of Health, Building 31, Bethesda, MD 20892.

NSF (National Science Foundation) Grant Policy Manual. A compendium of basic NSF policies and procedures for use by the grantee community and by NSF staff. Subscription service consists of a basic manual and supplementary material for an indeterminate period. Subscription: $13 (domestic); $16.25 (foreign). Superintendent of Documents, Government Printing Office, Washington, DC 20402.

NSF Bulletin. Provides monthly news about NSF programs, deadline dates, publications, meetings, and sources for more information. Issued monthly (except July and August) by the National Science Foundation, 1800 G Street, NW, Washington, DC 20550. Free.

The NSF publishes and updates at irregular intervals a publication listing fellowship opportunities: "A Selected List of Fellowship Opportunities and Aids to Advanced Education for United States Citizens and Foreign Nationals." (In 1987, the current list is dated Spring 1984.) The list may be requested from the National Science Foundation, The Publications Office, 1800 G Street, NW, Washington, DC 20550.

The following circulars are available from the Public Affairs Office, Executive Office of the President, Washington, DC 20500, and should be in the files of all nonprofit organizations who receive research support from the federal government:

OMB Circular A-21. Cost principles for educational institutions.

OMB Circular A-110. (Due to be revised in 1987.) Uniform administrative requirements for grants and agreements with institutions of higher education, hospitals, and other nonprofit organizations.

OMB Circular A-122. Cost principles for nonprofit organizations.

Foundations. *The Foundation Directory, Edition 10.* Describes more than 4,400 foundations, whose assets total more than $63 billion, and which award $4.1 billion in grants annually, accounting for 92 percent of all foundation awards in the United States in 1983 and 1984. (Edition 11 covering 1985 and 1986 will be published in the fall of 1987.) Each listing includes address, telephone number, financial information, purpose of the foundation, and grant application information. Published by the Foundation Center, 79 Fifth Avenue, New York, NY 10003. Toll-free number: 800-424-9836. $65.

National Data Book, 9th Edition. Lists all foundations (approximately 24,000) that give more than $1 per year; lists more than 200 community foundations that operate their own charitable or research programs. Gives addresses, assets, and has an index in a separate volume to aid in finding the number and size of small foundations in any community; also gives grant-making levels of foundations by state or region. The Foundation Center, 79 Fifth Avenue, New York, NY 10003. Toll-free number: 800-424-9836.

Taft Foundation Information System. A foundation directory, *Taft Foundation Reporter,* with information on more than 500 major private foundations, plus two monthly periodicals to update the directory data. Taft Corporation, 5125 MacArthur Boulevard, NW, Washington, DC 20016. $360/system; *Foundation Reporter* only, $287; subscription to monthly periodicals only, $97/year.

Corporations. Corporate foundations are included in *The Foundation Directory* and *National Data Book.* There is no reliable guide to corporate giving, except for those funds that flow through the company foundations. Corporations support research primarily in institutions with special expertise or facilities that do not exist in the corporate laboratories and that might otherwise have to be acquired for one research project. Research contracts with corporations usually grow out of professional relationships that have been built through collaborative work, and seldom go to strangers. The head of a research laboratory should know what corporations have branches or installations in the area and should be informed about the kinds of research they might have a need for. Approaching the research director of a corporation with an offer to provide some special facility or expertise is more likely to meet with success than sending any number of letters of application.

Grant Applications

Application forms are used by government agencies, and they change from time to time, so it is important to obtain the appropriate, up-to-date form from the agency to which an application is to be sent.

Many foundations now also use forms, but the first approach to a foundation or a corporation is best made by sending a letter describing the proposed work and asking for an interview to discuss it.

If it is possible to visit and talk with the program officer in a government agency, a foundation, or a corporation, by all means do so, but be prepared to answer questions about the research plan, time required, personnel and facilities required, and an estimate of the cost. No approach is as effective as the face-to-face discussion with an official who may become your advocate within the granting agency when the application is being considered.

Even if there has been extensive preliminary discussion, there always comes a time when a formal, written application must be submitted. Every research scientist should know how to prepare a grant application. NIH applications come with very detailed instructions for their presentation, but it may be helpful to refer to samples of successful proposals. *Grant Proposals That Succeeded*, published by Plenum Press, in 1983, contains samples of proposals funded by NSF, NIMH, and a very detailed illustration of an NIH grant application, prepared by George N. Eaves, David R. Schubert, and Steven C. Bernard.

Other useful references for proposal writing are:

1. Jagger, J. How to Write a Research Proposal. *Grants Magazine* 3(4):216–222, 1980.
2. DeBakey, L. and DeBakey, S. The Art of Persuasion: Logic and Language in Proposal Writing. *Grants Magazine* 1(1):45–60, 1978.
3. DeBakey, L. The Persuasive Proposal. *Journal of Technical Writing and Communication* 6:5–25, 1976.
4. Eaves, G.N. The Grant Application: An Exercise in Scientific Writing. *Federation Proceedings* 32:1541–1543, 1973.

Private foundations and corporations guard their application review process with great secrecy, and even insiders are not always clear about how the system works. One former foundation official noted, "Decisions frequently 'emerge' in the Quaker sense." The role of the program officer is highly significant, and often that individual serves as a "friend at court" to argue the case for the applicant. The personality, influence, and stature of the official in that role may be decisive, but unfortunately the applicant usually has no choice as to which official receives and processes an application. The best that can be done is to establish good rapport with the corporation or foundation official who is in charge of the program to which the application is submitted, arm that individual with information in addition to that contained in the formal application, which may give further support in favor of the project proposed, and hope that he or she will be eloquent in presenting your case. If the official is only lukewarm about the proposal, it does not augur well for approval but does not necessarily mean disapproval. Decisions by foundations and corporations are somewhat mysterious.

Government agencies dispense public funds, and feel a greater responsibility to be open about their selection process and to make awards on the basis of merit, with regard for national needs and some attention to geographic distribution.

Peer Review. Applications for governmental support are reviewed by staff members of the funding agency and by at least one review group selected for special knowledge in the field. In some cases, the final decision is made by a council or committee based upon the recommendations of reviewers, the program director, and the agency head; in other cases, it is the agency head who makes the final decision, based upon the recommendations of the advisory group.

The Public Health Service (PHS) of the Department of Health and Human Services (HHS) insists that all applications for support of biomedical research be subject to review by highly respected representatives of the scientific community. This is referred to as the "Peer Review System."

The major weakness in any review system is that panels and committees are composed of human beings who bring with them the usual complement of human failings and prejudices. They come with gaps in knowledge and flaws in judgment; some may favor older scientists, others may be biased toward the young. Some are impressed by famous institutions, and there are some who lean over backward to accommodate the up-and-coming or struggling ones. It is impossible to create a review system without shortcomings, but the peer review process used by the National Institutes of Health (NIH) of PHS is surely one of the best that has so far been devised. The system will be described here in some detail because such a large proportion of applications for governmental support are sent to that agency.

NIH applications are first read by a scientist and assigned to the appropriate institute; in case of overlaps in scientific subjects, the application may be sent to two institutes. Simultaneously, it is assigned to the appropriate study section (a peer review committee) for scientific appraisal. If there is no study section with the composition or expertise to handle the subject of the proposal, an ad hoc study section may be formed from among NIH consultants knowledgeable in the field. Each study section is served by an executive secretary who is responsible for coordinating and reporting the review of each application assigned to that section. The executive secretary determines which two or more members of the study section are best qualified to evaluate the proposal and sends it to those members by mail. The reviewers thus selected must prepare a detailed written critique of the proposals and return them to the secretary in advance of the meeting of the full study section. Formal study section meetings are held three times a year, and six to eight weeks before the meetings, all members receive copies of all proposals to be reviewed, usually 50 to 100, which each member is expected to read. Thus each member may read ten to twenty applications in detail, prepare a critique of each one, and read all the others.

The study section executive secretary is the intermediary between the applicant and the reviewers and if additional information or explanations are needed, the request is made through the executive secretary. If the applicant wishes to provide any additional material or communicate any information to the study section, that, too, must go through the executive secretary.

At the formal study section meetings each application is reported on by the assigned reviewers who prepared the detailed critique, after which other members may make comments or ask questions. A majority vote of the study section determines whether an application is approved, disapproved, or deferred for later consideration. Each approved application is given a numerical score, which is used to establish the priority rating transmitted to the National Advisory Council for each application. Immediately after the meeting of the study section, the executive secretary compiles the study section's discussion of an application into a summary statement that contains a detailed critique and the priority score assigned to the application. The critique is particularly important because it serves as a guide in preparing future applications, and it may develop that the unfunded application will be acceptable if revised and submitted for a later competition. As soon as each summary statement is completed, it is sent to the institute or other funding component, where the appropriate scientist administrator will read it and immediately send a copy to the principal investigator. Thus, the principal investigator will have some idea of the application's relative standing before it goes to the National Advisory Council for review.

Study-section review is based solely on scientific merit. Council review takes into account a broad area of considerations, including the needs of the NIH and the mission of individual institutes, the total pattern of research in nonprofit institutions, the need for initiation of research in new areas, the degree of relevance of the proposed research to the agency mission, and other matters. Councils do not change the priority scores of the study section but may recommend that an approved application be placed in a category to be funded or in one not to be funded based on program relevance or other policy considerations. Thus, the needs of the institute and the status and needs of the particular scientific area are taken into account, but the priority score still guides the final decision.

After council review, the professional staff in each institute takes over and matches the recommendations of the reviewers with the available funds. When it is not possible to fund all approved applications, a not unusual occurrence, the priority ranking of each proposal is critical. The professional staff member in charge of the program notifies the applicant of the final decision and all questions concerning the application from then on should be directed to the program director, including requests for information regarding the decision. Notification is usually made within a few weeks after the council meeting.

Applications that are approved, but which do not have a high enough priority rating for the funds currently available, are categorized and reported as "approved, but not funded." This means that if funds later become available there is a possibility that the proposal might be funded. This happens rather infrequently and, generally speaking, one should not be very hopeful.

NIH study sections meet three times a year, eight to ten weeks after the deadline for submission of applications. Council meetings may be three to four months following that. Consequently, the shortest time for the complete procedure is about nine months from submission of the application to notification of council action on funding. The safest course is to plan a project to start at least a year after the application is submitted and hope to have the decision within nine months.

The review of contract proposals at the NIH differs somewhat from the review of grant applications. When the availability of Requests for Proposals is announced, potential contractors may ask for the complete *Request for Proposal* (RFP), which contains specific instructions for preparation of the application, or to be more exact, the "contract offer."

Offers are submitted directly to the contracting officer, named in the RFP, by the established deadline date. They are first subjected to a review for technical evaluation. This technical merit review (TMR) is conducted by one or more panels in the scientific or technical discipline associated with the contract requirements. Reviewers may be government or nongovernment personnel depending upon the particular expertise needed; they are usually nongovernment experts.

Proposals found acceptable from the technical point of view are then reviewed from the standpoint of the elements of cost. The financial review, in addition to assessing the overall reasonableness of costs reflected in a proposal, may also disclose desirable shifts in emphasis that will affect the overall funding, such as changes in manpower loading or changes in material estimates. The contracting officer then enters into negotiations with all those who offer proposals within an acceptable technical and financial range for each project. In a few cases, especially for large contracts—those that will be $500,000 or more—negotiation may be done by means of a site visit. For smaller contracts, negotiations will be conducted by telephone. During such discussions, the offerer has an opportunity to present his position with respect to any aspect of the contract requirements.

After discussions or site visits, the contracting officer will make a decision based upon the offer that seems to hold the greatest technical advantages to the government, cost and other factors considered. More than one award may be made under the same RFP. Successful offerers are notified as soon as the decision is made, and written notice is sent to unsuccessful offerers that their proposals were not accepted.

The contract seldom (or never) comes as a surprise. In the negotiations dealing with cost factors or other points, it soon becomes clear that the contracting officer is viewing the proposal with favor. In some cases, a revised budget, which determines the award total, is worked out in detail and agreed upon in discussion.

Those who propose contracts may receive, upon request, an explanation or evaluation of the considerations that resulted in the failure of a proposal to be accepted. This can be a useful learning experience, providing excellent counsel and guidance for future proposals.

The decision-making process for contracts used to take less time than for grants, and occasionally it still does, but it is best to be prepared for a wait of several months.

Independent government agencies also have review systems that involve more than one type of evaluation and more than one panel. The National Science Foundation (NSF) programs, like the NIH ones, are administered by highly qualified professionals in the fields for which grants and contracts are available. Staff review of applications is the major selection device; however, the program personnel at NSF are assisted in the evaluation process by other scientists selected especially for their knowledge of the proposal subject.

There are actually two forms of peer review or "merit review," as it is now called, at NSF. In the biological, social, and behavioral sciences programs, the program officers are assisted by advisory panels that meet regularly two or three times a year to review all proposals in the program to which they are advisors and to make recommendations as to which are worthy of support. Program officers then decide on the basis of funds available and on their own judgment which of the applications approved by the panel will receive awards.

Physical, mathematics, and engineering program officers select ad hoc reviewers who are knowledgeable in the field and, after receiving their recommendations, make the selection of those to be funded. This system is similar to that used by military departments that make research grants.

The Site Visit. Projects involving large sums of money, or long-range support, are frequently site visited. Opinions vary on the value of such visits. If the facilities of an institution are well known, if the principal investigator is sufficiently prominent, and if the agency has previously supported projects in the same location involving the same personnel, it may seem unnecessary to take up the time of the panel and the applicant institution for such a visit. But, in some cases, visits are absolutely necessary if reviewers are to do justice to the proposal. It may be, for example, that certain facilities are essential to carry out the work properly and the reviewers may want to make sure that not only are the facilities in place at the institution but that they will be available at the time they are required for the performance of the work. Site visits are also an excellent way to ascertain the willingness of the institution to cooperate in carrying out the work. There are many informal, nonspecific ways in which the administration of an institution can encourage and support an activity or do just the opposite. The role of the institutional head in the site visit is to ensure that the institution will cooperate fully in the execution of the project.

A well-handled site visit can go far toward creating good will for the institution and for the proposed project. The institution should extend its normal hospitality to the visitors, but lavish entertainment of site visitors tends to make them suspicious. Representatives of federal agencies traveling at government expense may permit the host institution to provide them with lunch. After all, site visitors are "guests," in the technical sense of the word, and to offer them lunch is a normal gesture of courtesy and a convenience for all.

Grants Administration

Grant awards from the federal government come with detailed information on their management. This is not always the case with grants from corporations or foundations. It is, however, imperative that the principal investigator and the institutional accounting officer be informed in detail concerning the management and use of external funds. One of the most injudicious errors that can occur in an institution is the comingling of internal and external funds. It is difficult to believe that this occurs anywhere now but, just as a precaution, it must be noted that this is the road to sure disaster. The institution may find it necessary to advance funds for start-up of a new project when an award has been made but funds have not yet arrived. As long as the internal funds are available and the advance is carefully documented, this is not only acceptable but it is essential in some cases. However, this is where a red flag goes up. Grant applicants and contract offerers who have been in communication with the contracts and grants officer at the awarding agency can easily interpret a remark to mean that the award has been approved when it is only in the final and seemingly favorable stage. No internal funds should ever be advanced to start up a grant or contract project until official notification of the award has been received.

As soon as the funds arrive, an account must be set up for each project and authorized signatures established for the expenditure of the funds. The principal investigator, or someone designated by the principal investigator, should be the signatory for the disbursement of any funds in the account. But the institutional accounting office is responsible to ascertain that all disbursements are appropriate within the terms of the grant award. Scientists are not always the best business managers, and, occasionally, their own philosophy can conflict with the guidelines for expenditures under a particular award. Make no mistake about it. The authority for expenditure of any external funds is the award document, and if there is any question about interpretation, the final arbiter is the awarding agency.

All government funding agencies and many foundations and corporations expect the grantee institution to send a final financial report and a report of the work done under the grant. For long-term awards, requiring the submission of continuation requests, progress reports must accompany the application for continued funding. Some federal agencies can be extremely hard-nosed about these reports and will refuse to give final approval on grants to an institution, in some cases, until overdue reports have been filed.

Financial statements may be prepared in the accounting department or, if there is an office for grants administration, they may be handled by that office, but the principal investigator, who authorizes all expenditures, should be a party to the preparation or at least review it and understand it before signing it.

Progress reports are prepared by the principal investigator, and the director of research or someone delegated the task should review every progress report and final report that goes to the grantor. This is not only a safeguard against the reporting of misleading, erroneous, or potentially embarrassing data; it is also one way for the head of the institution to keep abreast of the research being done and to assess the achievements of the research staff.

Beginning in March 1986, the National Institutes of Health, the National Science Foundation, the Office of Naval Research, and the Departments of Energy and Agriculture are conducting an experimental project to cut bureaucratic red tape and reduce the role of federal agencies in grant management. This is to be done by turning over most routine project management tasks to the grantee institution, including authority to approve line-item changes in a project's budget and one-time program extensions of up to one year. The system foresees the necessity to strengthen the institutional management of federal grants. If the experiment proves successful, as it promises to do, its supporters hope the revised Circular A-110 scheduled for 1987 will reflect the changes in regulations governing administration of federal grant funds.

MAKING BUDGET CUTS

"Fair is a four-letter word!" said Lord Flowers, vice-chancellor of the University of London, when, as Sir Brian Flowers, he chaired the British Science Research Council. At one time labeled the enfant terrible of British science, Lord Flowers was dramatizing his belief that research funds should be distributed on the basis of quality without regard for demographic, democratic, compassionate, or other considerations.

Leland Haworth, one of the most respected scientist-administrators the United States ever produced, agreed with this principle. "Faced with the necessity for making severe budget cuts," said Haworth, "the first impulse of some administrators is to say, 'We've been cut 15 percent (or whatever), so it's 15 percent all around.' This is one of those crutches used by weak-minded administrators."

When retrenchment is inevitable, key research staff or faculty should be informed immediately. Invite their counsel, ask for their cooperation, and keep them informed at every step. Senior scientists do not like to learn that the place is in trouble or that their budgets have been cut by overhearing it in the cafeteria line, or by having their technicians bring it back from the coffee break.

Given the opportunity to participate in the cutback policies, department or lab heads will sometimes voluntarily suggest reductions within their own units in order to protect those activities they value most. In one research institution—rare in that it has a tenure policy—the tenured research staff volunteered to forego salary raises to help meet one budget cut.

Senior staff and key personnel can be particularly helpful in decisions regarding reductions in services and supplies. A common mistake of administrators, under pressure from higher-ups and in order to impress the staff with the seriousness of the situation, is to drastically cut amenities that result in wasted time and inconvenience to the researchers and bring about very little actual savings. Reduction of some services can seriously affect research activities to such an extent that the sacrifice is greater than the gain. Curtailments must be justified by the significance of the resultant savings, and the resulting disaccommodation and diminution in service should not be so great as to cripple a vital program.

After consulting with the staff and all voluntary reductions have been made, it is the head of the organization who must make the really hard, often painful, decisions.

The following ten points are suggested as a guide for administrators faced with the necessity for making sizable budget cuts that cannot be absorbed without some losses in programs and personnel:

1. Adopt the right attitude. Recession can be a useful adversity. It forces the establishment of priorities, the determination of what is truly indispensable. These priorities must be set objectively and honestly, based upon scientific merit in relation to the goals and research capability of the laboratory. Keep in mind that the most valuable research programs may require expensive services without which their work cannot be fully accomplished. The identification of an activity as a "service" should not automatically mark it for abolition.

2. The best scientific talent and most significant programs must be retained. Attrition cannot be expected to solve the problem; it often occurs in those critical programs where replacements are mandatory.

3. The morale of the institution will directly reflect that of the director. If the management message is "Things are terrible and you had better prepare for the worst," many scientists will take that as a cue to leave and the best ones will be the first to go—they will get the best offers. The positive leader says, "We are faced with a severe challenge which we can meet if we all work together."

4. Consult the senior staff. All cuts made must be acceptable to them *in principle*, which means they must understand them even though they hurt.

5. Once the priorities have been established, never deviate from them in order to accommodate a favorite program or a colleague of long stand-

ing. Morale will suffer if the criteria are strictly applied in some cases and relaxed in others.

6. Avoid making penny-wise and pound-foolish cuts, those that are more cosmetic than real. After selecting for retention the most valuable activities and the most talented staff, it is demoralizing for them to be subjected to petty economies that do not add up to significant savings.

7. Maintain some forward movement. At the very worst of times, some new activities must be introduced, some new appointments made. The staff will go along with this because they know that in the long run the lab will die without new ideas.

8. Be mindful of the human factors. Make every effort to transfer staff members within the organization if possible, and to assist others in finding positions.

9. Do not compromise on the standards for acceptance of projects sponsored by grants or contracts. It is tempting in time of recession to take on projects that would normally be considered incompatible with the quality and aspirations of the institution. The nature and stature of the laboratory can be completely changed if the availability of outside funds is allowed to determine the research program.

10. At the first sign of improvement in the situation, for example, a windfall from an unexpected source, let everyone know about it and participate in the decision about its use, if possible. And, of course, at the earliest stages of recovery, begin to restore salary cuts, services, and other necessary and desirable items that have been eliminated or reduced. Hardships that are borne with fortitude in bad times can arouse the mutinous instincts of the staff in better times.

Budget slashing is one of those benchmarks that measure the stature of the scientific administrator. Almost anyone can survive in prosperous times; mistakes made when funds are plentiful can be hidden or tolerated; some weak programs can be supported as long as they do not threaten those of the highest quality. It is during times of financial hardship that real leadership shines out like a beacon in a storm, when decisiveness, confidence, strength, and, above all, courage make the difference. As Lord Flowers cautions, being "fair"—that is, evenhanded—will not resolve the matter; neither will being afraid. "Fear" is a four-letter word, too.

Chapter 14

THE NEXT GENERATION: WHO WILL DO SCIENCE?

Let us now praise famous men'—
. . . For their work continueth,
. . . Greater than their knowing!
Rudyard Kipling, *Stalky & Co. A School Song*

Knowledge grows exponentially, and the remarkable advances made in the twentieth century may be seen as the pinnacle of a structure to which building blocks were added through centuries by scholars passing on their discoveries to disciples and pupils.

In the seventeenth century, Sir Isaac Newton wrote about those building blocks in a letter to Robert Hooke: "If I have seen further (than you and Descartes) it is by standing upon the shoulders of Giants."

Ever since young men eager for knowledge followed Socrates through the streets of Athens, and perhaps long before, sages understood that their wisdom owed much to what had been handed down to them and that they could achieve scholarly immortality only by providing the base upon which the next generation would add its contribution to the world's knowledge.

The time has long since passed, however, when a Socrates could hold discourse while strolling through the agora, when a Plato could set up a private Academy to expound his learning, or when an Aristotle, miffed at being passed over as head of the Academy when Plato died, could form his own Lyceum and surround himself with pupils. Modern education, and especially modern scientific education, requires extensive facilities and equipment and has become a public policy matter and an inescapable function of the modern research organization.

The head of a laboratory sees to the continuing education of the research staff, without which they risk falling behind in current developments, particularly in fast-moving fields. What's more, every institution has an obligation to foster the scholarly tradition of passing on to the next generation the accumulated wisdom of the present one, not only for the sake of tradition, but also because the laboratory's survival may depend upon it.

The education of scientists has, in fact, become a national obligation. Any nation that aspires to become preeminent in scientific discovery and to remain so cannot ignore its duty to seek out young talent and to encourage and support the entry of promising students into scientific careers.

Scientific advances cannot be bought with money; resources are necessary, but the limiting factor is talent. It has become a matter of deep concern that the American education system may not be producing, in the numbers required, the highly trained people needed to perform scientific research, to develop new technologies, and to teach the new generation. Poor schooling and low motivation threaten to confront us with a shortage of people trained in the quantitative disciplines—the biological, physical, and computer sciences, and mathematics and engineering.

Between 1970 and 1980, approximately 33,000 Ph.D.s were awarded annually; but in 1970, 45 percent of those were in science and by 1980, science accounted for only 33 percent. Some observers blame the socioeconomic climate; the generation that wants to "have it all" can get it faster in the business world and on Wall Street. But there is evidence that the main culprit is the educational system, as indicated by the depressing report of the National Commission on Excellence in Education, which concluded that, "If an unfriendly foreign power had attempted to impose on America the mediocre educational performance that exists today, we might well have viewed it as an act of war."

Blaming the educational system vents our anger, but it doesn't fully explain why the sciences no longer attract great numbers of the best and the brightest of the nation's young students. And it does not come to terms with the fact that much of the potential scientific talent is being wasted through neglect and discouragement, particularly among the minorities—blacks, Hispanics, American Indians, and women.

The scientific, academic, and philanthropic establishment has begun to deal with this issue to a limited extent and at a relatively late stage in the educational experience.

The National Science Foundation has announced an award for "Women, Minority, and Handicapped Engineering Research Assistants," which provides funds for undergraduates or high school students as research assistants on engineering projects. The support may be used for a summer, a quarter, or an academic year.

The National Institutes of Health has announced that in the evaluation of applications for the National Research Service Award (NRSA) Program for training in biomedical and behavioral research, special attention be given to "recruiting individuals from minority groups that now are underrepresented nationally in the biomedical and behavioral sciences."

Four universities—Stanford, University of California at Berkeley, Cornell, and Princeton—have entered into a joint arrangement under which minority undergraduates are paid stipends to travel to, and to study with research scientists at any of those universities over a summer. The Carnegie Corporation supports a major effort to promote mathematics education for women and minorities at Berkeley's Mathematics and Science Education Program for Women. The Ford Foundation makes grants to improve the mathematical competence of minority students. The Rockefeller Foundation is supporting programs to retain minority students in the science education pipeline in Houston, New York, and Pittsburgh.

These are laudable moves, but they may be too late to develop the best scientific talent from every available pool. In the study "Who Will Do Science?" commissioned by the Rockefeller Foundation in 1983, Sue E. Berryman of the Rand Corporation concluded that the pool of scientific talent can be identified as early as elementary school; that it reaches maximum size prior to high school and is essentially complete by grade 12. After high school, movement is almost entirely out of the pool.[1]

Research laboratories have long accepted responsibility for postdoctoral and some predoctoral training and consider those to be the appropriate levels for an institution devoted to full-time research. It may, at first, strain the imagination of senior scientists to envision themselves participating in the educational guidance of pre–high school students, to say nothing of devising a program for so doing. Scientists with young children are generally more receptive to the idea as they may be involved in activities where such guidance is encouraged, for example, Scouts or other youth organizations, young people's religious groups, hobby clubs, or in-school science projects. But the obligation to awaken and nurture scientific talent in the young goes beyond one's immediate family or neighbors; it means deeper involvement in the community beyond one's own ethnic, religious, professional, economic, or social circle.

For very young students, scientists may give classroom talks and simple demonstrations of scientific experiments, lead nature walks, organize hobby or science clubs, and so on. Summer jobs for students beginning at the high school level can often be funded by grants or other external sources. They provide students with a powerful incentive for continuing their interest in science, while inspiring them through an awareness of their own potential to strive for the highest achievement of which they are capable.

There is clear evidence that much of our scientific talent is never realized because it withers unrecognized and ignored in those very early years, and scientists who care about "Who Will Do Science?" have an obligation to help answer that question.

REFERENCES

1. Bruer, J.T. Who Will Do Science?—Minorities and Women If We Let Them. *RF Illustrated* December 1983, p. 15–16.

INDEX